INVITATION TO QUANTUM MECHANICS

INVITATION TO QUANTUM MECHANICS

Daniel F Styer
Oberlin College, USA

NEW JERSEY • LONDON • SINGAPORE • BEIJING • SHANGHAI • HONG KONG • TAIPEI • CHENNAI • TOKYO

Published by

World Scientific Publishing Co. Pte. Ltd.
5 Toh Tuck Link, Singapore 596224
USA office: 27 Warren Street, Suite 401-402, Hackensack, NJ 07601
UK office: 57 Shelton Street, Covent Garden, London WC2H 9HE

British Library Cataloguing-in-Publication Data
A catalogue record for this book is available from the British Library.

INVITATION TO QUANTUM MECHANICS

ISBN 978-981-124-790-3 (hardcover)
ISBN 978-981-124-928-0 (paperback)
ISBN 978-981-124-791-0 (ebook for institutions)
ISBN 978-981-124-792-7 (ebook for individuals)

For any available supplementary material, please visit
https://www.worldscientific.com/worldscibooks/10.1142/12576#t=suppl

Desk Editor: Nur Syarfeena Binte Mohd Fauzi

Typeset by Stallion Press
Email: enquiries@stallionpress.com

Love all God's creation, the whole and every grain of sand in it.
Love the stars, the trees, the thunderstorms, the atoms.
The more you love, the more you will grow curious.
The more you grow curious, the more you will question.
The more you question, the more you will uncover.
The more you uncover, the more you will love.
And so at last you will come to love the entire universe with an agile and resilient love founded upon facts and understanding.

— This improvisation by Dan Styer was inspired by the first sentence, which appears in Fyodor Dostoyevsky's *The Brothers Karamazov.*

Dedicated to *Linda Ong Styer*, adventurer

Contents

Synoptic Contents

5. Solving the Energy Eigenproblem

Since energy plays a central role, we devote a chapter to solving such problems. We find that solving particular problems strengthens our conceptual understanding, and that conceptual understanding strengthens our skill in solving particular problems.

6. Identical Particles

This surprisingly subtle topic deserves a chapter of its own.

7. Atoms

We apply our new knowledge to physical (rather than model) systems.

8. The Vistas Open to Us

This book is an invitation. Where might you and quantum mechanics travel together?

Welcome

Why would anyone want to study quantum mechanics?

Starting in the year 1900, physicists exploring the newly discovered atom found that the atomic world of electrons and protons is not just smaller than our familiar world of trees, balls, and automobiles, it is also fundamentally different in character. Objects in the atomic world obey different rules from those obeyed by a tossed ball or an orbiting planet. These atomic rules are so different from the familiar rules of everyday physics, so counterintuitive and unexpected, that it took more than 25 years of intense research to uncover them.

But it is really only since the year 1990 that physicists have come to appreciate that the rules of the atomic world (now called "quantum mechanics") are not just *different* from the everyday rules (now called "classical mechanics"). The atomic rules are also far *richer*. The atomic rules provide for phenomena like particle interference and entanglement that are simply absent from the everyday world. Every phenomenon of classical mechanics is also present in quantum mechanics, but the quantum world provides for many additional phenomena.

Here's an analogy: Some films are in black-and-white and some are in color. It does not malign any black-and-white film to say that a color film has more possibilities, more richness. In fact, black-and-white films are simply one category of color films, because black and white are both colors. Anyone moving from the world of only black-and-white to the world of color is opening up the door to a new world — a world ripe with new possibilities and new expression — without closing the door to the old world.

This same flood of richness and freshness comes from entering the quantum world. It is a difficult world to enter, because we humans have no experience, no intuition, no expectations about this world. Even our language, invented by people living in the everyday world, has no words for the new quantal phenomena — just as a language among a race of the color-blind would have no word for "red".

Reading this book is not easy: it is like a color-blind student learning about color from a color-blind teacher. The book is just one long argument, building up the structure of a world that we can explore not through touch or through sight or through scent, but only through logic. Those willing to follow and to challenge the logic, to open their minds to a new world, will find themselves richly rewarded.

The place of quantum mechanics in nature

Quantum mechanics is the framework for describing and analyzing small things, like atoms and nuclei. Quantum mechanics also applies to big things, like baseballs and galaxies, but when applied to big things, certain approximations become legitimate: taken together, these are called the *classical approximation* to quantum mechanics, and the result is the familiar classical mechanics.

Quantum mechanics is not only less familiar and less intuitive than classical mechanics; it is also harder than classical mechanics. So whenever the classical approximation is sufficiently accurate, we would be foolish not to use it. This leads some to develop the misimpression that quantum mechanics applies to small things, while classical mechanics applies to big things. No. Quantum mechanics applies to all sizes, but classical mechanics is a good approximation to quantum mechanics when it is applied to big things.

For what size is the classical approximation good enough? That depends on the accuracy desired. The higher the accuracy demanded, the more situations will require full quantal treatment rather than approximate classical treatment. But as a rule of thumb, something as big as a DNA strand is almost always treated classically, not quantum mechanically.

This situation is analogous to the relationship between relativistic mechanics and classical mechanics. Relativity applies always, but classical mechanics is a good approximation to relativistic mechanics when applied to slow things (that is, with speeds much less than light speed c). The speed at which the classical approximation becomes legitimate depends upon the accuracy demanded, but as a rule of thumb particles moving less than a quarter of light speed are treated classically.

The difference between the quantal case and the relativistic case is that while relativistic mechanics is less familiar, less comforting, and less expected than classical mechanics, it is no more intricate than classical mechanics. Quantum mechanics, in contrast, is less familiar, less comforting, less expected, *and* more intricate than classical mechanics. This intricacy makes quantum mechanics harder than classical mechanics, yes, but also richer, more textured, more nuanced. Whether to curse or celebrate this intricacy is your choice.

speed

c

	small	big
fast	relativistic quantum mechanics	relativistic mechanics
slow	quantum mechanics	classical mechanics

0 — size

Finally, is there a framework that applies to situations that are both fast and small? There is: it is called "relativistic quantum mechanics" and is closely related to "quantum field theory". Ordinary non-relativistic quantum mechanics is a good approximation for relativistic quantum mechanics when applied to slow things. Relativistic mechanics is a good approximation for relativistic quantum mechanics when applied to big things. And classical mechanics is a good approximation for relativistic quantum mechanics when applied to big, slow things.

What you can expect from this book

This book introduces quantum mechanics at the second-year American undergraduate level. It assumes the reader knows about classical forces, potential energy functions, and the simple harmonic oscillator. The reader should know that wavelength is represented by λ, frequency by f, and that for a wave moving at speed c, $\lambda f = c$. S/he needs to know the meaning and significance of "standard deviation". Turning to mathematics, it assumes the reader knows about complex numbers (see appendix C) and dot products, knows the difference between an ordinary and a partial derivative, and can solve simple ordinary differential equations. It assumes that the reader understands phrases like "orthonormal basis representation of a position vector".

This is a book about physics, not mathematics. The word "physics" derives from the Greek word for "nature", so the emphasis lies in nature, not in the mathematics we use to describe nature. Thus the book starts with experiments about nature, then builds mathematical machinery to describe nature, and finally applies the machinery to atoms, where the understanding of both nature and machinery is deepened.

The book never abandons its focus on nature. It provides a balanced, interwoven treatment of concepts, techniques, and applications so that each thread reinforces the other. There are many problems at many levels of difficulty, but no problem is there just for "make-work": each has a "moral to the story". Some problems are essential to the logical development of the subject: these are labeled (unsurprisingly) "essential". Other problems promote learning far better than simple reading can: these are labeled "recommended". Sample problems build both mathematical technique and physical insight.

The book does not merely convey correct ideas, it also refutes misconceptions. Just to get started, I list the most important and most pernicious misconceptions about quantum mechanics: (*a*) An electron has a position but you don't know what it is. (*b*) The only states are energy states. (*c*) The wavefunction $\psi(\vec{x}, t)$ is "out there" in space and you could reach out and touch it if only your fingers were sufficiently sensitive.

I do *not* provide summary lists of key ideas and difficult-to-remember concepts and equations. That's because an equation that I find easy to remember might be hard for you to remember. I recommend instead that you write out for yourself, in your own words, a summary of the ideas and equations that *you* consider most important and most difficult to remember.

The object of the biographical footnotes in this book is twofold: First, to present the briefest of outlines of the subject's historical development, lest anyone get the misimpression that quantum mechanics arose fully formed, like Aphrodite from sea foam. Second, to show that the founders of quantum mechanics were not inaccessible giants, but people with foibles and strengths, with interests both inside and outside of physics, just like you and me.

Teaching tips

Most physics departments offer a second-year course titled MODERN PHYSICS. The topics in this course vary widely from institution to institution: special relativity and elementary quantum mechanics are staples, but the course might also cover classical waves, thermodynamics and elementary statistical mechanics, descriptive atomic, molecular, and solid state physics. No textbook could cover all this variety, nor should any textbook try: instead each institution should provide a mix of topics appropriate for its own students. This book is devoted only to quantum mechanics at the level of a MODERN PHYSICS course. You will want to add it to other materials for the other topics in your own particular course.

In chapters 2 and 3, a surprising amount of student difficulty comes from nothing more than getting straight which Stern-Gerlach analyzer is oriented in which direction. I recommend that you mark up some cardboard boxes to look like analyzers and analyzer loops, and use them as demonstrations during your classes.

Chapter 5 presents two techniques for solving the energy eigenproblem: one informal and one numerical. I discuss the first in class and assign the second for reading, because the first benefits from a lot of blackboard sketching, erasing, resketching, and gesturing. But your own priorities might differ from mine, so you might take the opposite tack.

This text spends a lot of time on concepts before applying those concepts to atoms. Atoms are mathematically intense, and it pays to get the concepts straight first. If we jumped directly into atoms, that mathematical intensity would completely obscure the conceptual issues. Some people like it that way, because they don't want to face the conceptual issues.

Acknowledgments

I learned quantum mechanics from stellar teachers. My high school chemistry teacher Frank Dugan introduced me not only to quantum mechanics but to the precept that science involves hard, fulfilling work in addition to dreams and imagination. When I was an undergraduate, John Boccio helped mold my understanding of quantum mechanics, and also molded the shape of my life. In graduate school N. David Mermin, Vinay Ambegaokar, Neil Ashcroft, and Michael Peskin pushed me without mercy but pushed me in the direction of understanding and away from the mind-numbing attitude of "shut up and calculate". My debt to my thesis adviser, Michael Fisher, is incalculable. I've been inspired by research lectures from Tony Leggett, Jürg Fröhlich, Jennifer and Lincoln Chayes, Shelly Goldstein, and Chris Fuchs, among others.

I have taught quantum mechanics to thousands of students from the general audience level through advanced undergraduates. Their questions, confusions, triumphs, and despairs have infused my own understanding of the discipline. I cannot name them all, but I would be remiss if I did not thank my former students Paul Kimoto, Gail Welsh, K. Tabetha Hole, Gary Felder, Sarah Clemmens, Dahyeon Lee, and Noah Morris.

This book has been in slow yet consistent development since 2010, and many students have given me feedback over these years. In the fall 2020 and spring 2021 semesters I taught Modern Physics at Oberlin College using a draft of this textbook. I received helpful corrections and suggestions from several students, but especially from Ilana Meisler. Thank you.

My scientific prose style was developed by Michael Fisher and N. David Mermin. In particular this book's structure of "first lay out the phenomena (chapters 1 and 2), then build mathematical tools to describe those phenomena" echos the structure of Fisher's 1964 essay "The Nature of Critical Points". I have also absorbed lessons in writing from John McPhee, Maurice Forrester, and Terry Tempest Williams. My teaching style has been influenced especially by Mark Heald, Tony French, Edwin Taylor, Arnold Arons, and Robert H. Romer.

David Kaiser corrected some of my misconceptions concerning history. Shelley Kronzek, my editor at World Scientific, kept faith in this project through multiple delays. The book was skillfully copyedited by Matthew Abbate (who was also my undergraduate roommate and who served as a potential guardian to my children).

Chapter 1

"Something Isn't Quite Right"

We are used to things that vary continuously: An oven can take on any temperature, a recipe might call for any quantity of flour, a child can grow to a range of heights. If I told you that an oven might take on the temperature of 172.1 °C or 181.7 °C, but that a temperature of 173.8 °C was physically impossible, you would laugh in my face.

So you can imagine the surprise of physicists on 14 December 1900, when Max Planck announced that certain features of blackbody radiation (that is, of light in thermal equilibrium) could be explained by assuming that the energy of the light could *not* take on any value, but only certain discrete values. Specifically, Planck found that light of frequency f could take on only the energies of

$$E = nhf, \qquad \text{where } n = 0, 1, 2, 3, \ldots, \tag{1.1}$$

and where the constant h (now called the "Planck constant") is

$$h = 6.626\,070\,15 \times 10^{-34} \text{ J s}. \tag{1.2}$$

That is, light of frequency f can have an energy of $3.0\,hf$, and it can have an energy of $4.0\,hf$, but it is physically impossible for this light to have an energy of $3.8\,hf$. Any numerical quantity that can take on only discrete values like this is called "quantized". By contrast, a numerical quantity that can take on any value is called "continuous".

The photoelectric effect (section 1.2) supplies additional evidence that the energy of light comes only in discrete values. And if the energy of light comes in discrete values, then it's a good guess that the energy of an atom comes in discrete values too. This good guess was confirmed through investigations of atomic spectra (where energy goes into or out of an atom via absorption or emission of light) and through the Franck–Hertz experiment (where energy goes into or out of an atom via collisions).

Furthermore, if the energy of an atom comes in discrete values, then it's a good guess that other properties of an atom — such as its magnetic moment — also take on only discrete values. The story of this chapter is that these good guesses have all proved to be correct.[1]

1.1 Light in thermal equilibrium: Blackbody radiation

You know that the logs of a campfire, or the coils of an electric stove, glow orange. You might not know that objects at higher temperatures glow white, although blacksmiths and glass blowers are quite familiar with this fact and use it to judge the temperature of the metal or the molten glass they work with. Objects at still higher temperatures, like the star Sirius, glow blue. A nuclear bomb explosion glows with x-rays.

Going down the temperature scale, the tables, chairs, walls, trees, and books around us glow with infrared radiation. (Many people are unaware of this fact because our eyes can't detect infrared light.) In fact, our own bodies glow in the infrared — at a somewhat shorter wavelength than our books, because our bodies are slightly warmer than our books. And the bitter cold of outer space glows with the famous 3 K cosmic microwave background radiation.

All these situations are examples of electromagnetic radiation — light — in thermal equilibrium. What does that mean? The light streaming from, say, a red neon tube is *not* in thermal equilibrium: for one thing, it has only one color, for another all the light streams in the same direction. Just as a stream of nitrogen molecules, each one with the same speed and each one moving in the same direction, is not in thermal equilibrium, so the red light, all the same wavelength and all moving in the same direction, is not in thermal equilibrium. But after that light is absorbed by matter at a given temperature, then re-emitted, then reabsorbed and then re-emitted again, several times, that light relaxes into equilibrium at the same

[1]This book is about physics, not the history of physics. In order to present the physical ideas clearly they are sometimes developed ahistorically. For example in the next section Planck's 1900 radiation law is developed as a refinement of the 1905 Rayleigh–Jeans law. In section 1.5.2 Bohr's 1913 atomic theory is developed as a consequence of de Broglie's 1924 concept of matter waves.

temperature as the matter with which it's been interacting. This light in thermal equilibrium has a variety of wavelengths, and it moves with equal probability in any direction. In exactly the same way, a stream of nitrogen molecules will, after many collisions, have a variety of molecular speeds, and the molecules move with equal probability in any direction.

If you want to do a high-accuracy experiment with light in thermal equilibrium, you will want light that has been absorbed and re-emitted many times. The worst possible object for putting light into thermal equilibrium would be a mirror, which reflects rather than absorbs most incoming light. Somewhat better would be matter painted white, which reflects much incoming light. Better still would be matter painted black. Best of all would be a cavity surrounded by matter, like a cave, so that the light in the cavity is absorbed by the walls and re-emitted many times. For these experimental reasons, light in thermal equilibrium is often called "blackbody radiation" or "cavity radiation".

Qualitative arguments explain a number of familiar features of blackbody radiation. You know that the atoms in matter oscillate: as the temperature increases, the oscillations become both farther and faster. The "farther" oscillations suggest that high-temperature objects should glow brighter; the "faster" oscillations suggest that they should glow with higher-frequency (hence shorter-wavelength) light. Turning these qualitative arguments into a quantitative prediction requires an understanding of electrodynamics and of statistical mechanics beyond the needs of this book. Here I summarize the reasoning involved. First, three principles from electrodynamics:

(1) In all cases, the state of light within a cavity can be expressed as a sum over the "normal modes" of light within that cavity. Normal modes come about when one half-wavelength fits within a cubical cavity; or two half-wavelengths; or three; or any integer. This principle states that if $\vec{E}_n(\vec{x},t)$ denotes the electric field due to the light of the normal mode indexed n, then the electric field of an arbitrary state of light is

$$\vec{E}(\vec{x},t) = \sum_n a_n \vec{E}_n(\vec{x},t) \tag{1.3}$$

where a_n sets the amplitude of that mode in the particular state. Each mode is characterized by a particular wavelength.

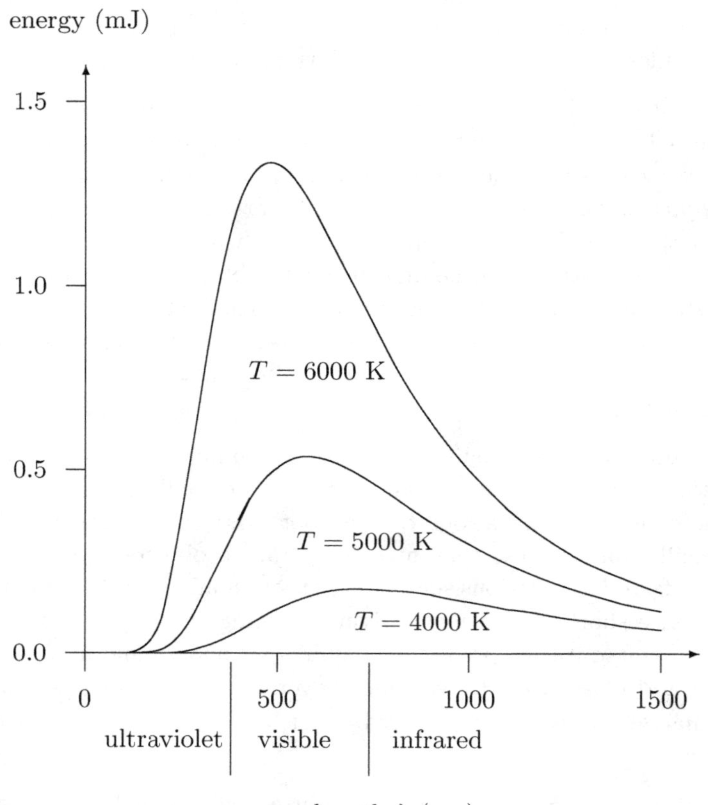

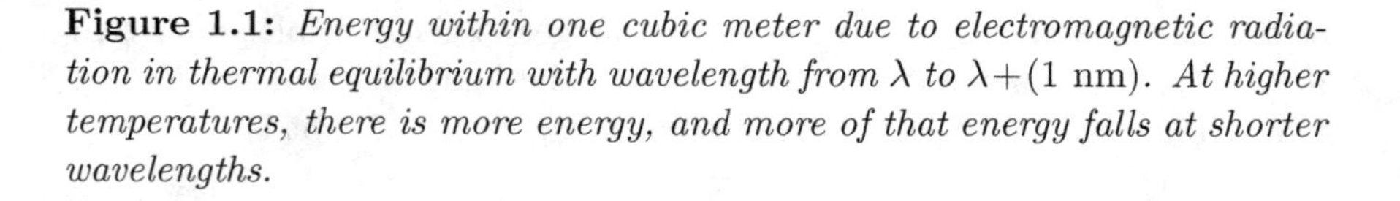
Figure 1.1: *Energy within one cubic meter due to electromagnetic radiation in thermal equilibrium with wavelength from* λ *to* $\lambda+(1 \text{ nm})$. *At higher temperatures, there is more energy, and more of that energy falls at shorter wavelengths.*

(2) The energy of any arbitrary state of light is just the sum of the energies due to each normal mode.

(3) There are more normal modes at shorter wavelength. In May 1905, Lord Rayleigh[2] calculated that the number of modes with wavelength falling within the window between λ and $\lambda + d\lambda$ was

$$\frac{64\,\pi V}{\lambda^4} d\lambda,$$

where V is the volume of the cavity (assuming a cavity large in the sense that $V \gg \lambda^3$). Two months later James Jeans[3] pointed out that Rayleigh had made a counting error: the correct result is

$$\frac{8\pi V}{\lambda^4} d\lambda. \tag{1.4}$$

To these three electrodynamics principles, add one principle from classical statistical mechanics: In thermal equilibrium, the energy[4] of each mode is $k_B T$, where T is the (absolute) temperature and k_B is the Boltzmann constant. This principle is called "equipartition": some modes have short wavelengths and some have long, but the energy is equally partitioned among the various different modes, independent of wavelength.

⟦The Boltzmann[5] constant k_B comes up whenever drawing a connection between energy and temperature. The U.S. National Institute of Standards and Technology gives its value as

$$k_B = 1.380\,649 \times 10^{-23} \text{ J/K}, \tag{1.5}$$

[2]John William Strutt, the third Baron Rayleigh (1842–1919), of England, is usually called just "Lord Rayleigh". Although particularly interested in acoustics, he made contributions throughout physics: he was the first to explain why the sky is blue and how seabirds soar. The mode counting described here was published in *Nature* on 18 May 1905.

[3]English physicist and astronomer (1877–1946). His 1930 popular book *The Mysterious Universe* did much to introduce the new quantum mechanics to a wide audience. His correct mode counting argument was published in *Philosophical Magazine* in July 1905. Rayleigh acknowledged his error ("Mr. Jeans has just pointed out that I have introduced a redundant factor 8.... I hasten to admit the justice of this correction.") in *Nature* on 13 July 1905.

[4]Technically the "average energy", but in these situations the thermal fluctuations about average energy are so small I'll just call it the "energy".

[5]Ludwig Boltzmann (1844–1906) of Austria developed statistical mechanics, which explains the properties of matter in bulk (such as density, hardness, luster, viscosity, and resistivity) in terms of the properties of atoms (such as mass, charge, and potential energy of interaction). His students include Paul Ehrenfest, who we will meet later, and Lise Meitner, who discovered nuclear fission.

but this number is hard to remember. For one thing, the joule (J) is a huge unit for measuring atomic energies — using it would be like measuring the distance between screw threads in miles or kilometers. The energy unit typically used in atomic discussions is instead the "electron volt", where

$$1 \text{ eV} = 1.60 \times 10^{-19} \text{ J}. \tag{1.6}$$

Second, I like to remember k_B through a product: at room temperature, about 300 K, the value of $k_B T$ is close to $\frac{1}{40}$ eV. (This knowledge has rescued me several times during physics oral exams.) The famous ideal gas constant R (as in $pV = nRT$) is just k_B times the Avogadro number, 6.02×10^{23}.]]

Putting these four principles together results in the "Rayleigh–Jeans law", which says that for light in thermal equilibrium at temperature T within a volume V, the electromagnetic energy due to wavelengths from λ to $\lambda + d\lambda$ is

$$(k_B T)\frac{8\pi V}{\lambda^4}d\lambda. \tag{1.7}$$

This formula has numerous admirable features: It is dimensionally consistent, as required. Doubling the volume results in doubling the energy, as expected. Higher temperature results in higher energy, in agreement with our previous expectation that "high-temperature objects should glow brighter". The very long wavelength modes are unimportant because there's no significant amount of energy in them anyway, so we can disregard our previous qualifier that the formula holds only when $V \gg \lambda^3$.

On the other hand, there is nothing in this formula supporting our expectation and common experience that high-temperature objects should glow with shorter-wavelength light. To the contrary, at *any* temperature the light spectrum should have exactly the same $1/\lambda^4$ character! Even worse: What is the total energy in blackbody radiation? It is

$$(k_B T)8\pi V \int_0^\infty \frac{1}{\lambda^4}d\lambda = (k_B T)8\pi V \left(-\frac{1}{3}\right)\left[\frac{1}{\lambda^3}\right]_0^\infty = +\infty. \tag{1.8}$$

Infinite energy! In fact, infinite energy at any finite temperature! If this were true, then every book and table and wall — not to mention every person — would be as deadly as an exploding nuclear bomb. The infinite energy arises from the short wavelengths of the spectrum, so this disastrous feature of the Rayleigh–Jeans prediction is called the "ultraviolet catastrophe".

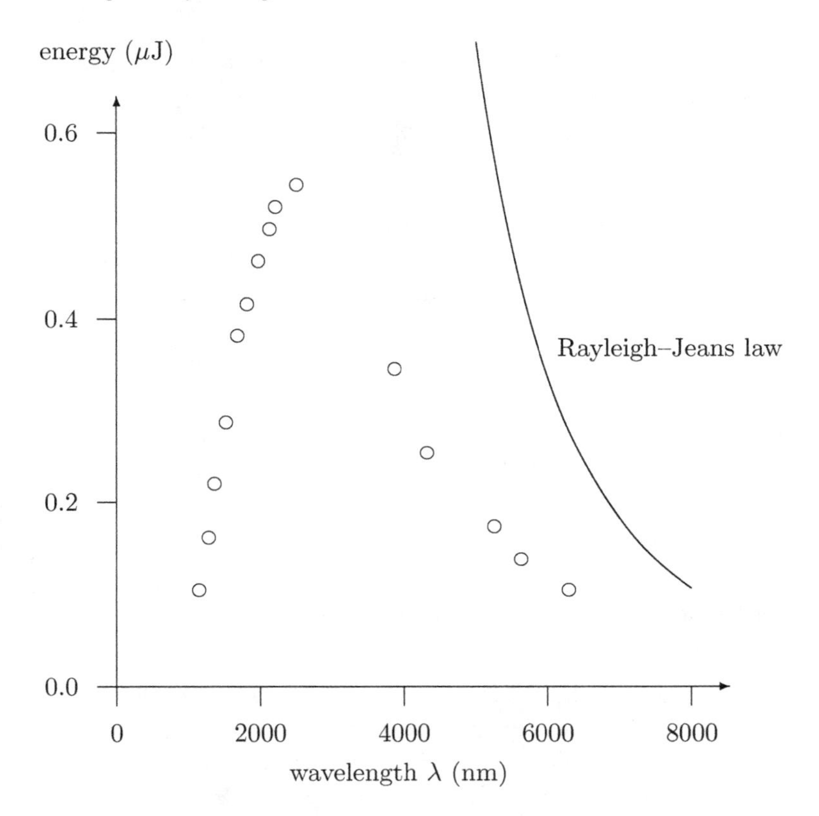

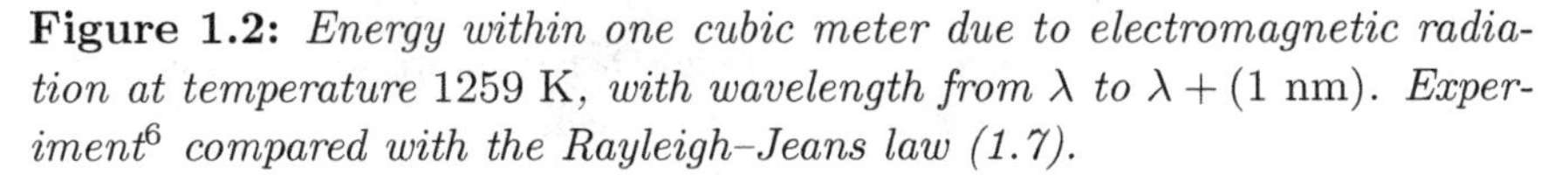
Figure 1.2: *Energy within one cubic meter due to electromagnetic radiation at temperature* 1259 K, *with wavelength from* λ *to* $\lambda + (1\ \mathrm{nm})$. *Experiment*[6] *compared with the Rayleigh–Jeans law (1.7).*

Experiment, and common experience, differ from the Rayleigh–Jeans law. However the experimental results mimic the Rayleigh–Jeans prediction at long wavelengths, so it's not totally off base. What could be wrong with the derivation? Perhaps the short-wavelength radiation is not really at thermal equilibrium. Perhaps there's some error in the mode counting that is significant only at short wavelengths. Or perhaps there's something wrong with the equipartition result.

[6]O. Lummer and E. Pringsheim, *Verhandlungen der Deutschen Physikalischen Gesellschaft* **2** (1900) 163.

In the year 1900 Max Planck[7] decided to pursue this third possibility. He wondered what would happen if the energy of a normal mode with frequency f couldn't take on any possible energy, but only certain values

$$E = nhf, \qquad \text{where } n = 0, 1, 2, 3, \ldots \tag{1.9}$$

and where h is some constant to be determined by a fit to experiment. (Except that, being a formal German professor, Planck didn't say that he "wondered", he said that he "hypothesized".) Because $f\lambda = c$ for light, Planck's hypothesis is equivalent to

$$E = n\frac{hc}{\lambda}. \tag{1.10}$$

When Planck worked out the statistical mechanical consequences of his hypothesis, he found that if it were correct then the energy of a mode would not be the equipartition result $k_B T$, but instead

$$\frac{hc/\lambda}{e^{(hc/\lambda)/(k_B T)} - 1}. \tag{1.11}$$

No longer would the energy be equally partitioned...instead it *would* depend upon wavelength.

Before rushing into the laboratory to test Planck's idea, let's see if it even makes sense. Planck's hypothesis is that the energy is discrete, not continuous, that it comes in packets of size hf. (Except that he didn't call them packets, he called them "quanta", from the Latin word for "amount". The singular is "quantum", the plural is "quanta".) When the frequency is small, that is, when the wavelength is long, these quanta are so small that the continuous approximation ought to be excellent. What does Planck's formula say for for long wavelengths? If λ is large, then $(hc/\lambda)/(k_B T)$ is small. How does e^x behave when x is small? Remember Taylor's formula:

$$e^x \approx 1 + x \qquad \text{when } |x| \ll 1.$$

Thus for long wavelengths

$$e^{(hc/\lambda)/(k_B T)} \approx 1 + \frac{hc/\lambda}{k_B T} \qquad \text{when } \lambda \gg \frac{hc}{k_B T}.$$

[7]Max Karl Ernst Ludwig Planck (1858–1947) was a German theoretical physicist particularly interested in thermodynamics and radiation. Concerning his greatest discovery, the introduction of quantization into physics, he wrote, "I can characterize the whole procedure as an act of desperation, since, by nature I am peaceable and opposed to doubtful adventures." [Letter from Planck to R.W. Wood, 7 October 1931, quoted in J. Mehra and H. Rechenberg, *The Historical Development of Quantum Theory* (Springer–Verlag, New York, 1982) volume 1, page 49.]

It follows that, in this same long-wavelength regime,

$$e^{(hc/\lambda)/(k_BT)} - 1 \approx \frac{hc/\lambda}{k_BT}$$

and

$$\frac{hc/\lambda}{e^{(hc/\lambda)/(k_BT)} - 1} \approx \frac{hc/\lambda}{(hc/\lambda)/(k_BT)} = k_BT.$$

The λs have canceled and we have recovered the equipartition result!

This analysis tells us two things: First, we know that for long wavelengths the Planck result (1.11) is almost the same as the equipartition result. Second, we know that the boundary between long and short wavelengths falls near the crossover wavelength

$$\lambda_\times = \frac{hc}{k_BT}. \tag{1.12}$$

This doesn't mean that Planck's formula is right, but it's not *transparently* wrong.

Adding Planck's result for energy to the same normal mode count result (1.4) that we used before results in the "Planck radiation law", which says that for light in thermal equilibrium at temperature T within a volume V, the electromagnetic energy due to wavelengths from λ to $\lambda + d\lambda$ is

$$\left(\frac{hc/\lambda}{e^{(hc/\lambda)/(k_BT)} - 1}\right)\frac{8\pi V}{\lambda^4}d\lambda. \tag{1.13}$$

This formula has the same admirable features possessed by the Rayleigh–Jeans result and discussed immediately below equation (1.7). (Does the electromagnetic energy increase with increasing temperature? See sample problem 1.1.2 on page 20.)

We're still not quite ready for the laboratory. Does Planck's result suffer from the same ultraviolet catastrophe that the Rayleigh–Jeans result did? This question is investigated in sample problem 1.1.1 on page 18. The result is: No, it doesn't.

Now is a good time to go to the laboratory. The Planck radiation law fits the data extraordinarily well, provided that one uses the value for h given in equation (1.2). Planck's hypothesis that an electromagnetic normal mode can't take on any energy, only certain values given by equation (1.9), seems to be correct.

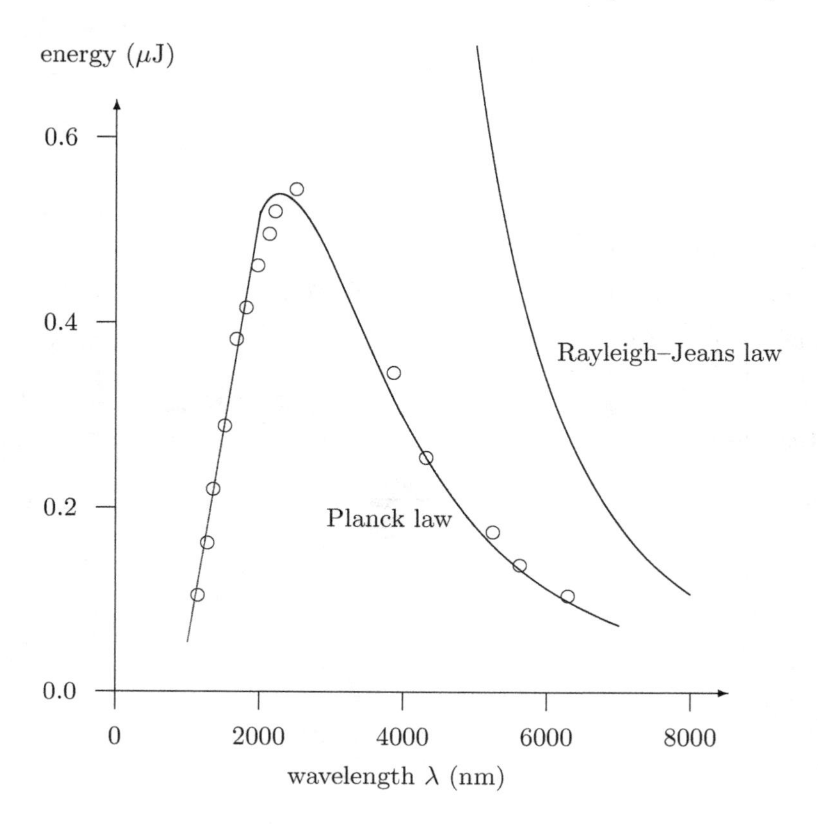

Figure 1.3: *Energy within one cubic meter due to electromagnetic radiation at temperature* 1259 K, *with wavelength from* λ *to* $\lambda + (1\text{ nm})$. *Experiment and Rayleigh–Jeans law compared with the Planck law (1.13).*

1.1.1 *Sample Problem: Stefan–Boltzmann law*

We saw that the Rayleigh–Jeans formula (1.7) could not represent reality, because it said that any body in thermal equilibrium would contain an infinite amount of radiant energy. Does the Planck formula (1.13) suffer from the same excruciating defect?

Solution: If the Planck radiation law holds, then the energy in blackbody radiation of *all* wavelengths is of course

$$\int_0^\infty \left(\frac{hc/\lambda}{e^{hc/(k_B T\lambda)}-1}\right)\frac{8\pi V}{\lambda^4}d\lambda. \tag{1.14}$$

You will be tempted to jump immediately into evaluating this integral, but I urge you to pause for a moment and find a good strategy before executing it. Integration is a mathematical, not a physical, operation, so I will first convert to a mathematical variable — that is, to a variable without dimensions. A glance suggests that the proper dimensionless variable is

$$x = \frac{hc/\lambda}{k_B T}. \tag{1.15}$$

(Comparison to the crossover wavelength $\lambda_\times$ in definition (1.12) shows that this variable is just the ratio $x = \lambda_\times/\lambda$.) Converting the integral to this variable shows that the total energy is

$$8\pi V\frac{(k_B T)^4}{(hc)^3}\int_0^\infty \frac{x^3}{e^x-1}dx. \tag{1.16}$$

Even without evaluating the integral, this equation gives us a lot of information. The integral on the right is just a number (unless it diverges) independent of T (or h, or c), so the total energy of light in thermal equilibrium is proportional to T^4. This fact, called the Stefan–Boltzmann law, is the formal result corresponding to our common experience that objects glow brighter at higher temperatures.

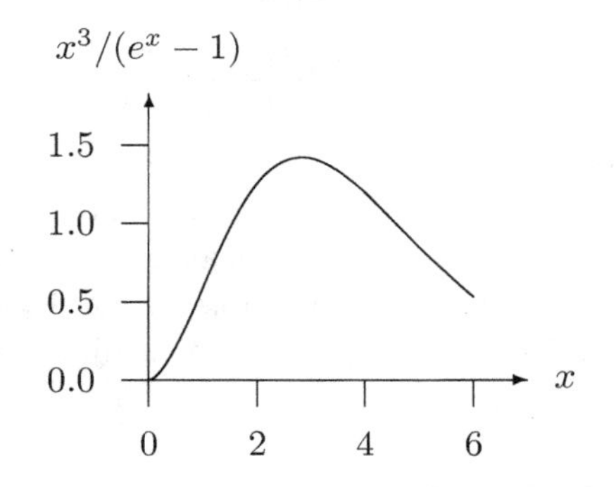

Figure 1.4: *Plot of the integrand $x^3/(e^x - 1)$. At small x (that is $x \ll 1$) the integrand behaves like x^2, at large x like $x^3 e^{-x}$.*

Now that we've squeezed the physics out, it's time to execute the math. You can work out the integral yourself, or you can use a computer algebra system, and in either case you'll find that the integral evaluates to $\pi^4/15$. The total energy is therefore

$$\frac{8\pi^5}{15} V \frac{(k_B T)^4}{(hc)^3}. \tag{1.17}$$

which is safely finite.

1.1.2 *Sample Problem: Characteristics of the Planck Radiation Law*

Your class has decided to write an "Underground Guide to Quantum Mechanics" for the benefit of next year's students in this course. You have volunteered to contribute about 500 words explicating the Planck radiation law (1.13). What do you write?

Solution: Like most equations in physics, the Planck radiation law (1.13 in our textbook) is concise, but this conciseness hides a lot of information. This essay unpacks some of this information to see what the equation is trying to tell us about nature. I address two points:

(1) Does the electromagnetic energy increase with increasing temperature?

There are a lot of symbols in the equation, but the temperature dependence comes in through only one term: The energy at temperature T is proportional to

$$\frac{1}{e^{T_0/T}-1}$$

where I have defined the temperature-independent constant $T_0 = hc/\lambda k_B$. As T increases, T_0/T decreases, so $e^{T_0/T}$ decreases, so $e^{T_0/T}-1$ decreases, so $1/(e^{T_0/T}-1)$ increases. Thus the electromagnetic energy increases with increasing T, as expected.

[Note, however, that energy doesn't increase *linearly* with temperature. You might know that for a monatomic gas modeled as noninteracting classical point particles, the total energy for N particles in thermal equlibrium at temperature T is $\frac{3}{2}Nk_BT$. From this, some people get the mistaken notion that energy and temperature are always related linearly, or even that this is the *definition* of temperature. Blackbody radiation provides a neat counterexample.]

If you like calculus, you could find the same result by showing that the slope of this curve is always positive. That slope is

$$\begin{aligned}
&\frac{d}{dT}\left(\frac{1}{e^{T_0/T}-1}\right)\\
&= -\frac{1}{(e^{T_0/T}-1)^2}\frac{d(e^{T_0/T})}{dT}\\
&= -\frac{1}{(e^{T_0/T}-1)^2}\frac{d(e^{T_0/T})}{d(1/T)}\frac{d(1/T)}{dT}\\
&= -\frac{1}{(e^{T_0/T}-1)^2}T_0e^{T_0/T}\left(-\frac{1}{T^2}\right)\\
&= \frac{e^{T_0/T}}{(e^{T_0/T}-1)^2}\frac{T_0}{T^2}
\end{aligned}$$

which sure enough is always positive.

(2) What does the Planck law say about the energy in very short wavelengths?

The graph on page 12 of our textbook shows very small energies at small wavelengths. In fact all three energy curves hide behind the axis!

Having investigated the behavior of energy with temperature, what can we say about the behavior of energy with wavelength?

Write the energy equation to emphasize the λ dependence using the constant crossover wavelength $\lambda_\times$ defined in equation (1.12) in the textbook. The energy is proportional to

$$\frac{\lambda_\times/\lambda}{e^{\lambda_\times/\lambda} - 1}.$$

As $\lambda \to 0$ this approaches ∞/∞. You could resolve this indeterminate form using l'Hôpital's rule, but it's more insightful to look at the rate of approach to infinity. Define $x = \lambda_\times/\lambda$, and investigate the behavior as $x \to \infty$. Both the numerator x and the denominator $e^x - 1 \approx e^x$ approach infinity, but e^x goes to infinity faster than x, faster than x^2, faster than x^3, even faster than $x^{10^{23}}$. As $x \to \infty$, the quantity

$$\frac{x}{e^x - 1}$$

is not just small, it's "exponentially small", which explains why it hides behind the axis.

Conclusions: An equation is not an inert blob of symbols awaiting numbers to "plug in and chug through". An equation is a troubadour singing songs about nature. The songs are interesting only if you listen for them. This essay has listened to two of the notes sung by the Planck radiation law, equation (1.13).

Problems

1.1 **How big is an atom?**
How many times can a liter of water be cut in half until you're left with a single water molecule?

1.2 **Units for atomic-sized energies**
Physicists are fond of measuring typical atomic energies in "eV/atom", chemists are fond of measuring typical atomic energies in "kJ/mole". An energy of exactly 1 eV in an atom corresponds to what energy, in kJ, in a mole of atoms? (In reporting your numerical result, be sure to use significant figures and units: see appendices A and B.)

1.3 **Behavior of integrand**
Justify the claim, made in the caption of the graph on page 20, that when $x \ll 1$, $x^3/(e^x - 1) \approx x^2$. [*Clue:* $e^x = 1 + x + \frac{1}{2}x^2 + \cdots$.]

1.4 **Wien displacement law** (recommended problem)
Use the Planck radiation law (1.13) to show that at any given temperature, the wavelength holding maximum energy is $\hat{\lambda} = b/T$, where b is some constant that you don't need to determine. This so-called "Wien displacement law" is the formal result corresponding to our common experience that objects glow with shorter wavelengths at higher temperatures. [*Clue:* You could take the derivative of equation (1.13) with respect to λ, set the result equal to zero, and solve for $\hat{\lambda}$. This would be grotesquely difficult. Instead, using x as defined in equation (1.15), argue that the function $x^5/(e^x - 1)$ has a maximum located at, say, $\hat{x}$, and then derive an expression for $\hat{\lambda}$ in terms of $\hat{x}$ and other quantities. You don't need to find a numerical value for $\hat{x}$, just argue that it exists.]

1.2 Photoelectric effect

But does Planck's hypothesis apply only to light in thermal equilibrium? Does it apply only to normal modes? Here's the relevant question: What if we produce light of wavelength λ proceeding in a beam. This light is surely not in thermal equilibrium, nor is it in a normal mode. Will the energy in the beam still be restricted to the possible values

$$E = n\frac{hc}{\lambda} = nhf \qquad \text{where } n = 0, 1, 2, 3, \ldots? \tag{1.18}$$

In 1905 Albert Einstein[8] thought this question was worth pursuing. He reasoned in four steps:

(1) The electrons in a metal can be used as light-energy detectors.

(2) The reason the electrons are inside the metal at all, instead of roaming around free, is because it's energetically favorable for an electron to be within the metal. Some of the electrons will be bound tightly within

[8]Although Albert Einstein (1879–1955) is most famous for his work on relativity, he claimed that he had "thought a hundred times as much about the quantum problems as I have about general relativity theory." [Remark to Otto Stern, reported in Abraham Pais, *"Subtle Is the Lord...": The Science and the Life of Albert Einstein* (Oxford University Press, Oxford, UK, 1982) page 9.]

the metal, some less so. Call the trapping energy of the least-well-bound electron U_t.

(3) If an electron within the metal absorbs a certain amount of energy from the light shining on it, then some of that energy will go into getting the electron out of the metal. The rest will become kinetic energy of the ejected electron. The electron ejected with maximum kinetic energy will be the one with the minimum trapping energy, namely U_t. [The reasoning up to this point has been purely classical.]

(4) Add to this the quantum hypothesis: the amount of light energy available for absorption can't take on any old value — it has to be hf. Then the maximum kinetic energy of an ejected electron will depend upon the frequency of the light shining on the metal:

$$\mathrm{KE}_{\mathrm{max}} = hf - U_t. \tag{1.19}$$

If the light has low frequency, $hf < U_t$, then no electron will be ejected at all.

This analysis is clearly oversimplified. A metal is a complex system rather than a simple potential energy well, and it ignores the possibility that the electron might absorb an energy of $2hf$, or $3hf$, or more, but it's worth an experimental test: Plot the measured maximum kinetic energy of ejected electrons as a function of frequency f. Will the plot give a straight line with slope h matching the slope determined through the completely different blackbody radiation experiment?

The challenge was taken up by Robert A. Millikan.[9] By 1916 he experimentally verified Einstein's prediction, writing that "Einstein's photoelectric equation has been subjected to very searching tests and it appears in every case to predict exactly the observed results." Nevertheless, he found the quantum condition troubling: this same 1916 paper calls quantization a "bold, not to say reckless, hypothesis" because it "flies in the face of the thoroughly established facts of interference", and in his 1923 Nobel Prize acceptance speech he said that his confirmation of the Einstein equation had come "contrary to my own expectation".

[9]American experimental physicist (1868–1953), famous also for measuring the charge of the electron through his oil-drop experiment, and for his research into cosmic rays. A graduate of Oberlin College and of Columbia University, he used his administrative acumen to build the small vocational school called Throop College of Technology into the research and teaching powerhouse known today as the California Institute of Technology. His photoelectric results were reported in "A direct photoelectric determination of Planck's 'h'" *Physical Review* **7** (1 March 1916) 355–390.

Despite Millikan's reservations, the conclusion is clear: The energy of electromagnetic waves of frequency f cannot take on just any old value; it can take on only the values

$$nhf \qquad \text{where } n = 0, 1, 2, 3, \ldots.$$

This result is counterintuitive and non-classical, but it *is* in accord with blackbody and photoelectric experiments, and that's what matters.

One way to picture this energy restriction is to imagine the light as coming in noninteracting particles, each particle having energy hf. These pictured particles are called "photons". If no photons are present, the electromagnetic energy is 0; if one photon is present the energy is hf, if two photons are present the energy is $2hf$, and so forth. Using this picture, the blackbody and photoelectric experiments are said to demonstrate the "particle nature of light".

It is important to realize that this particle picture goes above and beyond what the experiments say. The experiments tell us that the energy can take on only certain values; they say nothing about particles. Because light, classically, consists of electric and magnetic fields, it is tempting to picture a photon as a "ball of light", a packet of classical electric and magnetic fields. We will see soon that the picture of a photon as a classical particle with a precise position, a precise energy (hf), *and* a precise speed (c) is not tenable. We will also encounter energy restrictions that cannot be interpreted through this picture at all: for example, the energy of a hydrogen atom is restricted to the values

$$-\frac{\text{Ry}}{n^2} \qquad \text{where } n = 1, 2, 3, \ldots$$

and where Ry is a constant. There is *no way* to picture this energy restriction through a collection of noninteracting hypothetical particles. The picture of photons can be made quite precise and can be very valuable, but only if you keep in mind that a photon does *not* behave exactly like a familiar classical point particle.

Further evidence for the quantized character of electromagnetic energy comes from the Compton effect (which involves the interaction of x-rays and electrons; see problem 1.9 on page 28), from the discrete clicks produced by a photomultiplier tube or any other highly sensitive detector of light energy, and from photon anticoincidence experiments. I will not describe these experiments in detail, but you should understand that the evidence for energy quantization in light is both wide and deep.[10] It is worth your effort to memorize that electromagnetic energy comes in lumps of magnitude

$$E = \frac{hc}{\lambda} = \frac{1240 \text{ eV}\cdot\text{nm}}{\lambda}. \tag{1.20}$$

⟦This equation comes up frequently in the physics Graduate Record Exam and in physics oral exams. I recommend that you remember the constant hc in terms of the unit usually used for energy in atomic situations, namely the electron volt, and the unit usually used for the wavelength of optical light, namely the nanometer.⟧

1.2.1 *Sample Problem: Find the flaw*

No one would write a computer program and call it finished without testing and debugging their first attempt. Yet some approach physics problem solving in exactly this way: they get to the equation that is "the solution", stop, and then head off to bed for some well-earned sleep without investigating whether the solution makes sense. This is a loss, because the real fun and interest in a problem comes not from our cleverness in finding "the solution", but from uncovering what that solution tells us about nature. (Appendix D, "Problem-Solving Tips and Techniques", calls this final step "reflection".) To give you experience in this reflection step, I've designed "find the flaw" problems in which you don't find the solution, you only test it. Here's an example.

This is a physics problem that you are *not* supposed to solve:

> Blackbody radiation is largely infrared at room temperature, largely red in a campfire, largely blue in the star Sirius. What is the relationship between the wavelength holding the peak energy, called $\hat{\lambda}$, and the temperature?

[10] A clear summary of the evidence that light energy is quantized, but that a photon is not just like a small, hard version of a classical marble, is presented in section 2.1, "Do photons exist?", of George Greenstein and Arthur G. Zajonc, *The Quantum Challenge* (Jones and Bartlett Publishers, Sudbury, Massachusetts, 2006). See also J.J. Thorn, M.S. Neel, V.W. Donato, G.S. Bergreen, R.E. Davies, and M. Beck, "Observing the quantum behavior of light in an undergraduate laboratory" *American Journal of Physics* **72** (September 2004) 1210–1219.

(This relationship enables astronomers to find the temperatures of distant stars inaccessible to human-made thermometers.)

Four friends work this problem independently. When they get together afterwards to compare results, they find that they have produced four different answers! Their candidate answers are

$$\text{(a)}\quad \hat{\lambda} = 0.201\frac{hc^2}{k_BT}$$

$$\text{(b)}\quad \hat{\lambda} = 0.201\frac{hc}{k_BT}$$

$$\text{(c)}\quad \hat{\lambda} = 0.201 \times 10^{-3}\frac{hc}{k_BT}$$

$$\text{(d)}\quad \hat{\lambda} = 0.201\frac{k_BT}{hc}$$

Provide simple reasons showing that three of these candidates must be wrong.

Solution: Candidate (a) does not have the correct dimensions for wavelength. There are no problems with candidate (b), which is in fact the correct "Wien displacement law". Candidate (c) claims that at room temperature ($k_BT = \frac{1}{40}$ eV), the dominant wavelength would be $\hat{\lambda} = 10$ nm, deep in the ultraviolet, whereas the problem statement tells you that it's actually in the infrared. (Recall that $hc = 1240$ eV·nm.) Candidate (d) not only has incorrect dimensions, it also shows the dominant wavelength increasing, not decreasing, with temperature.

Problems

1.5 **Visible light photons** (recommended problem)
The wavelength of visible light stretches from 700 nm (red) to 400 nm (violet). (Figures with one significant digit.) What is the energy range of visible photons?

1.6 **Light bulb photons**

Stand about 100 meters from a 60-watt light bulb and look at the bulb. The pupil in your eye has a diameter of about 2 mm. Estimate the number of photons entering one of your eyes each second. You will need to make reasonable assumptions: be sure to spell them out.

1.7 **Rephrasing the Einstein relation** (essential problem)

Using your knowledge of classical waves, rewrite the Einstein relation for the energy of a photon, $E = hc/\lambda$, in terms of the angular frequency $\omega = 2\pi f$ of light. Employ the shortcut notation $\hbar = h/2\pi$ and compare your result to

$$E = \hbar\omega. \tag{1.21}$$

1.8 **Character of photons**

Two classical particles, say two asteroids, interact with a gravitational potential energy, and each particle can have any possible non-negative kinetic energy. Write a paragraph or two contrasting these characteristics of classical particles with the characteristics of photons.

1.9 **Compton scattering**

(This problem requires background in relativity.)

When x-rays shine on a target, they are scattered in all directions. Both the wave and photon pictures of light predict this scattering. The wave picture, however, predicts that the scattered x-rays will have the *same* wavelength as that of the incoming waves, whereas the photon picture predicts that the wavelength of the scattered x-rays will depend upon the direction of scattering. This problem derives that dependence.

An x-ray photon of energy E_0 strikes a stationary electron of mass m. The photon scatters off with energy E at angle θ, the electron recoils with momentum p at angle ϕ. Recall from your study of relativity that, for both photon and electron, $E^2 - (pc)^2 = (mc^2)^2$, but that the photon mass is zero.

initial:

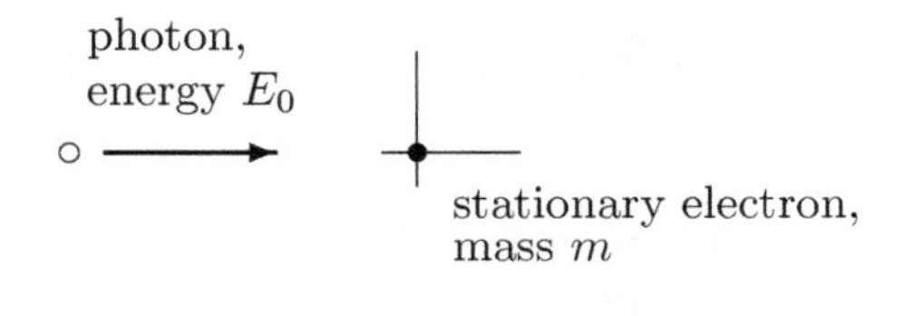

final:

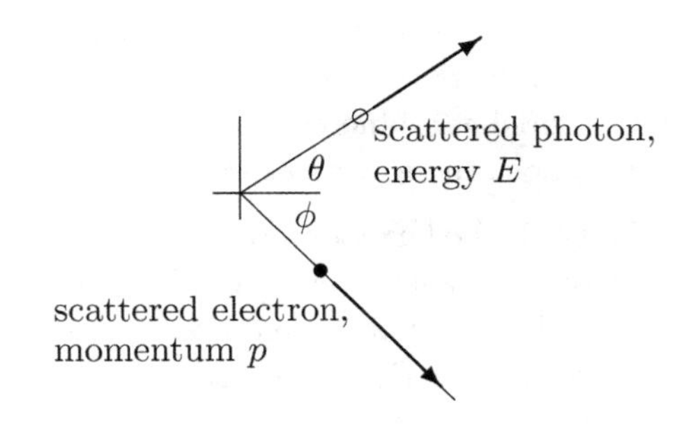

a. Write down the expressions for energy conservation, for momentum conservation in the horizontal direction, and for momentum conservation in the vertical direction, using only the variables in the figure above. Compare your result to

$$\begin{aligned} E_0 + mc^2 &= E + \sqrt{(mc^2)^2 + (pc)^2} \\ E_0/c &= (E/c)\cos\theta + p\cos\phi \\ 0 &= (E/c)\sin\theta - p\sin\phi. \end{aligned}$$

b. It is hard to detect electrons and (relatively) easy to detect x-rays. Hence we set out to eliminate the quantities involving the scattered electron, namely p and ϕ. There are three equations and we wish to eliminate two variables, so we expect to end up with one equation. Begin by squaring and combining the last two equations to eliminate ϕ, finding

$$E_0^2 - 2E_0E\cos\theta + E^2 = (pc)^2.$$

c. Meanwhile, show that the energy conservation equation is equivalent to

$$E_0^2 - 2E_0E + E^2 + 2(E_0 - E)mc^2 = (pc)^2.$$

Combine these two equations to find

$$(E_0 - E)mc^2 = E_0E(1 - \cos\theta).$$

d. So far, this has been a relativistic collision problem. It becomes a quantum mechanics problem when we note that, for a photon, $E = hc/\lambda$. Show that the x-ray wavelength changes by

$$\lambda - \lambda_0 = \frac{h}{mc}(1 - \cos\theta).$$

e. Let's read meaning into this equation to make it something more than a jumble of symbols. Does the photon picture predict an increase or decrease in wavelength? For what angle does the wave-picture prediction $\lambda = \lambda_0$ hold true?

In 1923, Arthur Compton verified that scattered x-rays have exactly this angle dependence. His experiment convinced many physicists that the photon picture — strange though it may be — must have some merit. Ever since then, a particle of mass m has been said to have a "Compton wavelength" of h/mc.

1.3 Wave character of electrons

If light, a classical wave, has some sort of particle character, could it be that an electron, a classical particle, has some sort of wave character?

The very idea seems absurd and meaningless: the word "particle" suggests a point, the word "wavelength" requires a length extending beyond a point. But in 1924 Louis de Broglie[11] thought the question was worth pursuing. He thought about the relationship between energy and wavelength for a photon:

$$E = \frac{hc}{\lambda}.$$

[11]Louis de Broglie (1892–1987) was born into the French nobility, and is sometimes called "Prince de Broglie", although I am told that he was actually a duke. He earned an undergraduate degree in history, but then switched into physics and introduced the concept of particle waves in his 1924 Ph.D. thesis.

De Broglie realized that if he wanted to extrapolate from a photon, which is intrinsically relativistic, to an electron, which might or might not be relativistic, he'd have have to use relativistic, not classical, mechanics. I hope you remember from your study of special relativity that for a photon — or anything else moving at light speed — the energy and momentum are related through $E = pc$. De Broglie stuck these two facts together and hypothesized that any particle has an associated wavelength given through

$$p = \frac{h}{\lambda}. \tag{1.22}$$

This associated wavelength is today called the de Broglie wavelength.

Please realize that this is not a derivation; it's more like a stab in the dark. And it's a difficult hypothesis to test, because particle wavelengths are typically so short. Recall from your study of classical waves that we test for wave character through interference experiments, and that such experiments work best when the apparatus is about the size of a wavelength. So what are the sizes of these de Broglie waves? In all cases $\lambda = h/p$, but in the non-relativistic case, a particle with kinetic energy K and mass m has $K = p^2/2m$, so $p = \sqrt{2Km}$ and thus the de Broglie wavelength is

$$\lambda = \frac{h}{\sqrt{2Km}} = \frac{hc}{\sqrt{2Kmc^2}}, \tag{1.23}$$

where in the last step I have inserted a factor of c in the numerator and the denominator to make the formula easier to remember.

⟦"What?" you object, "How can those extra cs, which just cancel out in the end, make anything easier?" First of all, I've already recommended on page 26 that you *not* memorize the value of h in SI units; instead you should remember that $hc = 1240$ eV·nm. Second, most people don't remember the mass of an electron m_e in kilograms; instead, they remember it through the energy equivalent $m_ec^2 = 0.511$ MeV. Similarly for all other elementary particles: I would have to look up the mass of a proton in kilograms, but I remember that $m_pc^2 \approx 1000$ MeV right off the top of my head.⟧

Let's try this out in a calculation. Suppose a stationary electron is accelerated through a potential difference of exactly 100 V, so it picks up a kinetic energy of $K = 100$ eV. That electron will have a de Broglie wavelength of

$$\begin{aligned}\lambda &= \frac{hc}{\sqrt{2Kmc^2}}\\ &= \frac{1240\text{ eV}\cdot\text{nm}}{\sqrt{2\times(100\text{ eV})\times(511\,000\text{ eV})}}\\ &= \frac{1240\text{ nm}}{\sqrt{2\times(10^2)\times(0.511\times10^6)}}\\ &= \frac{1240\text{ nm}}{\sqrt{1.022\times10^8}}\\ &= 0.123\text{ nm}.\end{aligned}$$

This is a very short wavelength — if it were an electromagnetic wave, it would be in the x-ray regime (10 nm to 0.01 nm). In 1924, it was impossible to manufacture slits — or anything else — with a size near 0.123 nm. (Higher-energy electrons, or protons of the same energy, would have even shorter wavelengths, and their wave character would be even harder to discern.)

This difficulty was overcome by using, not human-manufactured slits, but the rows of atoms within a crystal of nickel (with an atomic spacing of 0.352 nm). In a series of experiments executed from 1923 to 1927, Clinton Davisson and Lester Germer[12] showed that electrons scattering from nickel exhibit interference as predicted by de Broglie's strange hypothesis.

Since 1927, the wave character of particles, as demonstrated through interference experiments, has been tested time and again. One breakthrough came in 1987, when Akira Tonomura[13] and his colleagues at the Hitachi Advanced Research Laboratory in Tokyo demonstrated interference in electrons thrown one at a time through a classic two-slit apparatus. In 2013, Markus Arndt and his colleagues at the University of Vienna,[14] building on

[12]Davisson (1881–1958) and Germer (1896–1971) were experimenting at Bell Telephone Laboratories in Manhattan with the ultimate goal not of testing quantum mechanics, but of building better telephone amplifiers. Earlier, Germer had served as a World War I fighter pilot. Later, he would become an innovative rock climber.

[13]A. Tonomura, J. Endo, T. Matsuda, T. Kawasaki, and H. Ezawa, "Demonstration of single-electron buildup of an interference pattern" *American Journal of Physics* **57** (1989) 117–120. Tonomura (1942–2012) worked tirelessly to develop the full potential of electron microscopy, and almost as a sideline used these developments to test the fundamentals of quantum mechanics. If you search the Internet for "single electron two slit interference Akira Tonomura", you will probably find his stunning video showing how an electron interference pattern builds up over time.

[14]Sandra Eibenberger, Stefan Gerlich, Markus Arndt, Marcel Mayor, and Jens Tüxen,

work by Anton Zeilinger, demonstrated quantal interference in a molecule consisting of 810 atoms. Research in this direction continues.[15] Exactly what is meant by "wave character of a particle" needs to be elucidated, but the effect is so well explored experimentally that it cannot be denied.

Problems

1.10 **de Broglie wavelengths for various particles**
We have found the de Broglie wavelength of an electron with energy 100 eV. What about for a neutron with that energy? An atom of gold? (A neutron's mass is 1849 times the mass of an electron, and a gold nucleus consists of 197 protons and neutrons.) If these de Broglie waves had instead been electromagnetic waves, would such wavelengths be characterized as ultraviolet, x-rays, or gamma rays?

1.11 **de Broglie wavelength for a big molecule**
The 2013 experiments by Markus Arndt, mentioned in the text, used the molecule $C_{284}H_{190}F_{320}N_4S_{12}$ with mass 10 118 amu (where the "atomic mass unit" is very close to the mass of a hydrogen atom), and with a velocity of 85 m/s. What was its de Broglie wavelength?

1.12 **Rephrasing the de Broglie relation** (essential problem)
Rewrite the de Broglie relation, $p = h/\lambda$, in terms of the wavenumber $k = 2\pi/\lambda$. Employ the shortcut notation $\hbar = h/2\pi$ and compare your result to

$$p = \hbar k. \tag{1.24}$$

"Matter–wave interference of particles selected from a molecular library with masses exceeding 10 000 amu" *Physical Chemistry Chemical Physics* **15** (2013) 14696–14700.

[15] Jonas Wätzel, Andrew James Murray, and Jamal Berakdar, "Time-resolved buildup of two-slit-type interference from a single atom" *Physical Review A* **100** (12 July 2019) 013407.

1.13 de Broglie wavelength for relativistic particles

(This problem requires background in relativity.)

The kinetic energy of a relativistic particle with energy E and mass m is defined as $K = E - mc^2$. Recall that $E^2 - (pc)^2 = (mc^2)^2$.

a. Show that

$$(pc)^2 = (E - mc^2)(E + mc^2).$$

b. Combine the above result with the de Broglie relation $\lambda = h/p$ to show that the generalization of equation (1.23) applicable to both relativistic and non-relativistic particles is

$$\lambda = \frac{hc}{\sqrt{K(K + 2mc^2)}}.$$

c. Show that the above equation reduces to equation (1.23) in the non-relativistic limit. [*Clue:* Given two positive numbers, s and b, with $s \ll b$, then $s+b \approx b$, but there's no approximation for sb.]

1.14 Hofstadter's use of relativistic electrons

(This problem assumes you have completed the previous problem.)

The 1961 Nobel Prize in Physics was awarded to Robert Hofstadter "for his pioneering studies of electron scattering in atomic nuclei and for his thereby achieved discoveries concerning the structure of the nucleons". In one of these experiments he bombarded gold nuclei with electrons of total relativistic energy 183 MeV.

a. What is the de Broglie wavelength of such an electron?

b. Compare that wavelength to 10^{-15} m, the typical size of an atomic nucleus. (The experiment succeeded only because the wavelength of the electron probe was smaller than or comparable to the size of the nucleus. In exactly the same way, you can obtain a good image of a person's face using light — with a wavelength near 600 nm — but not using FM radio waves — with a wavelength near 3 m.)

c. Hofstadter reported his results in a paper[16] that didn't say whether the electron energy "183 MeV" refers to the total energy or just the kinetic energy. Is this a significant error on Hofstadter's part?

[16] R. Hofstadter, B. Hahn, A.W. Knudsen, and J.A. McIntyre, "High-energy electron scattering and nuclear structure determinations, II" *Physical Review* **95** (15 July 1954) 512–515.

1.4 How does an electron behave?

We have seen that an electron has a particle-like character, yet somehow it has a wave-like character as well. This seeming paradox invites the question: "How does an electron behave: like a particle or like a wave?"

I approach an answer through an analogy[17] drawn from another field of physics: the theory of classical waves. When a classical wave (water wave, sound wave, light wave, etc.) of wavelength λ passes through a slit of width a, wave theory tells us how the wave behaves: If the slit is large ($a \gg \lambda$; "geometrical optics limit") then the wave acts almost like a ray, which passes through the slit with no spreading. If the slit is small ($a \ll \lambda$; "spherical wave limit") then the wave acts like a Huygens wavelet, which passes through the slit and then spreads throughout the half-circle on the far side. For slits of intermediate size the wave acts in an intermediate manner. The behavior of a classical wave is known exactly and can be calculated with exquisite accuracy. Under some circumstances it behaves almost like a ray, and in some circumstances it behaves almost like a Huygens wavelet, although it takes on these behaviors exactly only in limiting cases.

As with classical waves, so with electrons. The theory of quantum mechanics tells us with exquisite accuracy how an electron behaves in all circumstances. Under some circumstances it behaves almost like a classical particle. Under other circumstances it behaves almost like a classical wave. The question "Does an electron behave like a classical particle or like a classical wave?" is like the question "Does a classical wave behave like a ray or like a Huygens wavelet?" It never behaves *exactly* like either. Instead, it behaves in its own inimitable way, which you might call "typical quantum mechanical behavior".

It is the job of this book (and of the rest of your physics education) to teach you "typical quantum mechanical behavior". If you open your mind to the idea that electrons behave in a manner unlike anything you have previously encountered, then you can gain an appreciation of and build an intuition for such "typical quantum mechanical behavior". If you refuse to admit this possibility, then you might be able to execute the problems, and you might even get a good grade, but your mind will be forever closed to one of the wonders of our universe.

[17]Perhaps you will find that the parable of "the blind men and the elephant" makes a more appealing analogy.

1.5 Quantization of atomic energies

If the energy of light comes in quantized amounts, how about the energy of an atom?

1.5.1 *Experiments*

It is not hard to change the energy of an atom: simply put a gas in a high-voltage discharge tube to give the gas atoms a jolt. The atoms will absorb energy from the discharge, and then will release light energy as they fall back to their ground state. If the atoms can take on continuous energy values, then they will emit light of continuous energy value and (through equation 1.20) of continuous wavelength. But if the atoms can take on only certain quantized energy values, then they will emit light of quantized energy value and hence of only certain wavelengths. Perhaps you have performed experiments sending the light from a discharge tube through a diffraction grating and spreading it out by wavelength: if so, then you know from your personal experience that the light comes out at only certain wavelengths, not with continuous wavelengths. If energy is released from an atom through light, it seems that the atom can take on only certain quantized energy values.

But what if the energy goes into or out of an atom through some other mechanism? In 1914 James Franck[18] and Gustav Hertz[19] figured out a way to get energy into mercury atoms through collisions with electrons. There's no need to go through the details of the Franck–Hertz experiment, but the conclusion is again that the atom cannot accept just any old amount of energy: it can only absorb energy in certain quantized amounts. Furthermore, those amounts agreed with the amounts derived from spectral experiments.

[18]German physicist (1882–1964) who left Germany in disgust after the Nazi Party came to power. He went first to Denmark, then to the United States where he worked to build the nuclear bomb. He authored a report recommending that U.S. nuclear bombs not be used on Japanese cities without warning.

[19]German physicist (1887–1975), nephew of Heinrich Hertz, for whom the unit of frequency is named. Hertz was forced out of his career in Germany due to distant Jewish ancestry. He went to the Soviet Union and there worked to build the Soviet nuclear bomb.

1.5.2 *Bohr's theory*

Niels Bohr[20] decided in 1913 not just to accept the quantization of atomic energies as an experimental fact, but to find a theoretical underpinning. He started with the simplest atom: hydrogen. Hydrogen consists of an electron (mass m_e, charge $-e$) and a far more massive proton (charge $+e$). Bohr made[21] two assumptions: first that the electron orbits the proton only in circular (never elliptical) orbits, second that the circumference of the circular orbit holds an whole number of de Broglie wavelengths (one or two or three or more but never 2.7). If you remember $F = ma$, and the formula for centripetal acceleration (v^2/r), and Coulomb's law, you'll realize that the circular orbit assumption demands

$$\frac{e^2}{4\pi\epsilon_0}\frac{1}{r^2} = m_e\frac{v^2}{r} = \frac{p^2}{m_e r}. \tag{1.25}$$

And if you remember the de Broglie formula $\lambda = h/p$ you'll realize that the assumption of a whole number of wavelengths demands

$$2\pi r = n\lambda = nh/p \qquad \text{where } n = 1, 2, 3, \ldots. \tag{1.26}$$

Here are two different formulas connecting the two variables r and p, so we can solve for these variables in terms of n, h, m_e, and $e^2/4\pi\epsilon_0$. (Don't get distracted by the quantities λ and v ... if we decide we want them later on we can easily find them once we've solved for r and p.) Once our objectives are clear it's not hard to achieve them. Solve equation (1.26) for p,

$$p = n\frac{h/2\pi}{r}, \tag{1.27}$$

and then plug this into equation (1.25) giving

$$\frac{e^2/4\pi\epsilon_0}{r^2} = n^2\frac{(h/2\pi)^2}{m_e r^3}.$$

Solve this equation for r giving

$$r = n^2\frac{(h/2\pi)^2}{m_e(e^2/4\pi\epsilon_0)}, \tag{1.28}$$

[20]Danish physicist (1885–1962), fond of revolutionary ideas. In 1924 and again in 1929 he suggested that the law of energy conservation be abandoned, but both suggestions proved to be on the wrong track. Father of six children, all boys, one of whom won the Nobel Prize in Physics and another of whom played in the 1948 Danish Olympic field hockey team.

[21]The treatment in this book captures the spirit but not all the nuance of Bohr's argument.

and then plug this back into (1.27) giving

$$p = \frac{1}{n}\frac{m_e(e^2/4\pi\epsilon_0)}{h/2\pi}. \tag{1.29}$$

(Notice that the quantity $e^2/4\pi\epsilon_0$ makes a natural combination of quantities, so we keep it together as a packet and never rend it apart in the course of our algebraic manipulations. Similarly for the quantity $h/2\pi$.)

If Bohr's assumptions hold, then the orbital radius can't be any old value, it can only take on the values given in (1.28). And the momentum can't be any old value, it can only take on the values given in (1.29). This is unexpected and curious and worthwhile, but in Bohr's day, and still today, our experimental apparatus is not so refined that it can determine the radius or the momentum of a single electron orbiting a proton, so it's also sort of useless. We can find an experimentally accessible result by calculating the energy

$$E = \frac{p^2}{2m_e} - \frac{e^2/4\pi\epsilon_0}{r}.$$

If Bohr's assumptions hold, then the energy, too, can't be any old value, it can only take on the quantized values

$$E = -\frac{1}{n^2}\frac{m_e(e^2/4\pi\epsilon_0)}{2(h/2\pi)^2}. \tag{1.30}$$

And, as discussed in section 1.5.1, the energy values of an atom *are* experimentally accessible.

Experiment shows this result for hydrogen to be correct. Yet the derivation clearly leaves much to be desired. The result (1.30) springs from two equations: equation (1.25) assumes that the electron is a classical point particle in a circular orbit; equation (1.26) assumes that the electron is a de Broglie wave. Both equations are needed to produce the energy quantization result, yet the two assumptions *cannot* both be correct.

1.5.3 *Visualizations*

As with the photon, it is the job of the rest of this book to come up with a description of an electron that is correct, but one thing is clear already: the visualization of an electron as a classical point particle — a smaller and harder version of a marble — cannot be correct. At this stage in your quantal education I cannot yet give you a perfect picture of an electron, but you can see that the picture of an electron as a point particle with a position, a speed, and an energy — the picture that appeared in, for example, the seal of the U.S. Atomic Energy Commission — must be wrong.

Figure 1.5: *An atom does not look like this.*

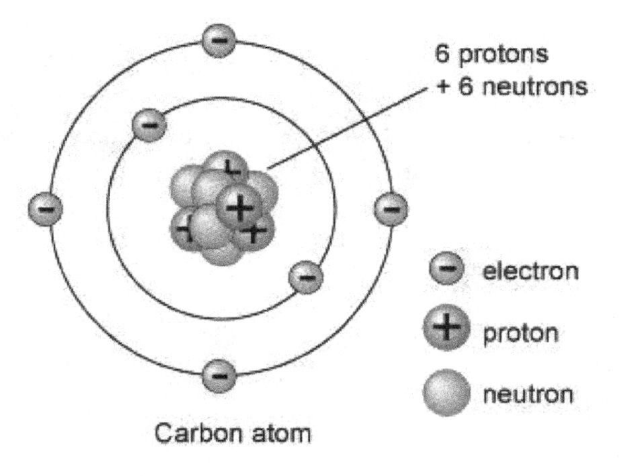

Figure 1.6: *An atom does not look like this, either.*

The electron is *not* a small, hard marble with a position, a speed, and an energy, and any intuition you hold in your mind based on that mistaken visualization is likely to lead you astray.

1.5.4 *Multi-electron atoms*

Bohr, of course, forged on to investigate atoms more complicated than hydrogen. He found that he could indeed explain the spectra of more complicated atoms — sometimes with high accuracy, sometimes only approximately — but *not* by making the assumption about circular orbits that worked so well for hydrogen. For example, here are the orbits required for the eleven electrons in a sodium atom:

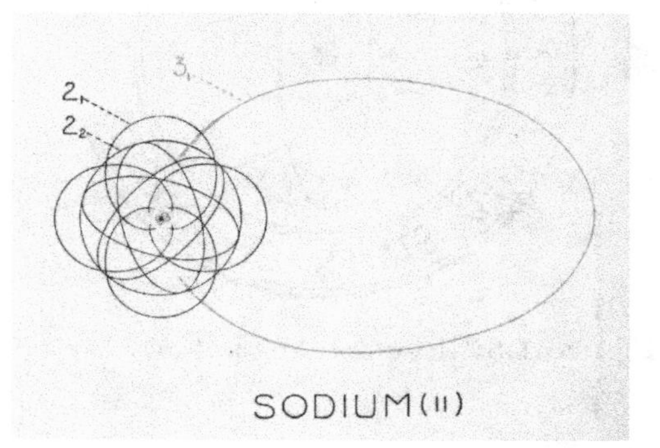

Figure 1.7: *A sodium atom in the Bohr model.*[22]

Atoms larger than sodium required even more elaborate schemes, and each new atom required a new set of assumptions. Furthermore, even for simple hydrogen, Bohr could never explain how the quantized energies changed when the atom was placed in a magnetic field (the "Zeeman[23] effect").

We must conclude that the Bohr model, despite its impressive prediction concerning the energies of a hydrogen atom in the absence of magnetic field, is wrong.

---

[22]Reproduced from K.A. Kramers and Helge Holst, *The Atom and the Bohr Theory of Its Structure* (Alfred A. Knopf, New York, 1923) rear endpaper.

[23]Pieter Zeeman (1865–1943), Dutch physicist, discovered this effect in 1896, three years after earning his Ph.D. The prestigious journal *Nature* had earlier described his observations of the meteoroid of 17 November 1882, made at age 17 years.

Problem

1.15 **Characteristic quantities** (recommended problem)

⟦It's a good idea to develop a sense of typical, or "characteristic", sizes: if a problem in classical mechanics asks you to calculate the mass of a squirrel, and you find 937 kg, then you know you've made a mistake somewhere. In classical mechanics this sense of typical quantities comes from everyday experience. In atomic physics this sense has to be built.[24] Although the Bohr model is not correct, it does provide a reasonable picture of typical sizes for atomic quantities, and this problem is your first step toward such a "tangible picture".⟧

The "characteristic length" for atomic systems is the so-called "Bohr radius", the radius of the smallest allowed orbit, which is (see equation 1.28)

$$a_0 \equiv \frac{(h/2\pi)^2}{m_e(e^2/4\pi\epsilon_0)}. \tag{1.31}$$

a. Evaluate the Bohr radius numerically in nanometers. Compare to a wavelength of blue light.

The "characteristic time" for atomic systems is conventionally defined not as the period of the smallest allowed Bohr orbit, but as this period divided by 2π. (Not the time for the electron to make one orbit, but the time for it to sweep out an angle of one radian.)

b. Derive a formula for this time and evaluate it numerically in femtoseconds. Compare to a period of blue light.

⟦Characteristic quantities for atomic systems will be explored further in problem 7.7, "Characteristic quantities for the Coulomb problem", on page 246. That problem shows why it makes sense to divide the period by 2π.⟧

[24] When the mathematician Stanislaw Ulam became interested in nuclear physics he "discovered that if one gets a feeling for no more than a dozen... nuclear constants, one can imagine the subatomic world almost tangibly, and manipulate the picture dimensionally and qualitatively, before calculating more precise relationships." [Stanislaw M. Ulam, *Adventures of a Mathematician* (Charles Scribner's Sons, New York, 1976) pages 147–148. (From the chapter "Life among the Physicists: Los Alamos".)]

1.6 Quantization of magnetic moment

An electric current flowing in a loop produces a magnetic moment, so it makes sense that the electron orbiting (or whatever it does) an atomic nucleus would produce a magnetic moment for that atom. And of course, it also makes sense that physicists would be itching to measure that magnetic moment.

It is not difficult to measure the magnetic moment of, say, a scout compass. Place the magnetic compass needle in a known magnetic field and measure the torque that acts to align the needle with the field. You will need to measure an angle and you might need to look up a formula in your magnetism textbook, but there is no fundamental difficulty.

Measuring the magnetic moment of an atom is a different matter. You can't even see an atom, so you can't watch it twist in a magnetic field like a compass needle. Furthermore, because the atom is very small, you expect the associated magnetic moment to be very small, and hence very hard to measure. The technical difficulties are immense.

These difficulties must have deterred but certainly did not stop Otto Stern and Walter Gerlach.[25] They realized that the twisting of a magnetic moment in a *uniform* magnetic field could not be observed for atomic-sized magnets, and also that the moment would experience zero net force. But they also realized that a magnetic moment in a *non-uniform* magnetic field *would* experience a net force, and that this force could be used to measure the magnetic moment.

[25]Otto Stern (1888–1969) was a Polish-German-Jewish physicist who made contributions to both theory and experiment. He left Germany for the United States in 1933 upon the Nazi ascension to power. Walter Gerlach (1889–1979) was a German experimental physicist. During the Second World War he led the physics section of the Reich Research Council and for a time directed the German effort to build a nuclear bomb.

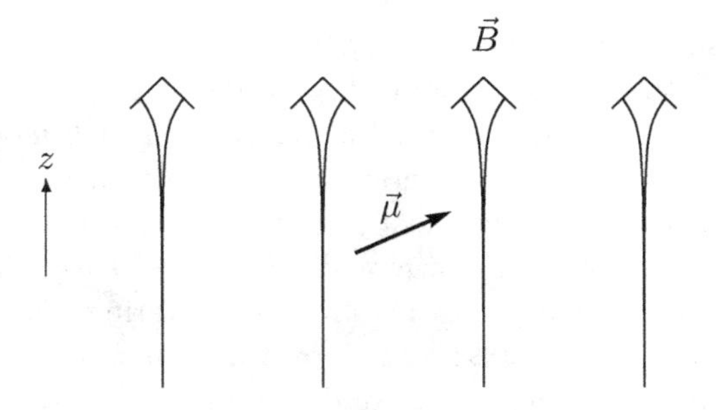

Figure 2.1: *A classical magnetic moment in non-uniform magnetic field.*

A classical magnetic moment $\vec{\mu}$, situated in a magnetic field $\vec{B}$ that points in the z direction and increases in magnitude in the z direction, is subject to a force

$$\mu_z \frac{\partial B}{\partial z}, \tag{1.32}$$

where μ_z is the z-component of the magnetic moment or, in other words, the projection of $\vec{\mu}$ on the z axis. (If this is not obvious to you, then work problem 2.16, "Force on a classical magnetic moment", on page 46.)

Stern and Gerlach used this fact to measure the z-component of the magnetic moment of an atom. First, they heated silver in an electric "oven". The vaporized silver atoms emerged from a pinhole in one side of the oven, and then passed through a non-uniform magnetic field. At the far side of the field the atoms struck and stuck to a glass plate. The entire apparatus had to be sealed within a good vacuum, so that collisions with nitrogen molecules would not push the silver atoms around. The deflection of an atom away from straight-line motion is proportional to the magnetic force, and hence proportional to the projection μ_z. In this ingenious way, Stern and Gerlach could measure the z-component of the magnetic moment of an atom even though any single atom is invisible.

Before turning the page, pause and think about what results you would expect from this experiment.

Here are the results that I expect: I expect that an atom which happens to enter the field with magnetic moment pointing straight up (in the z direction) will experience a large upward force. Hence it will move upward and stick high up on the glass-plate detector. I expect that an atom which happens to enter with magnetic moment pointing straight down (in the $-z$ direction) will experience a large downward force, and hence will stick far down on the glass plate. I expect that an atom entering with magnetic moment tilted upward, but not straight upward, will move upward but not as far up as the straight-up atoms, and the mirror image for an atom entering with magnetic moment tilted downward. I expect that an atom entering with horizontal magnetic moment will experience a net force of zero, so it will pass through the non-uniform field undeflected.

Furthermore, I expect that when a silver atom emerges from the oven source, its magnetic moment will be oriented randomly — as likely to point in one direction as in any other. There is only one way to point straight up, so I expect that very few atoms will stick high on the glass plate. There are many ways to point horizontally, so I expect many atoms to pass through undeflected. There is only one way to point straight down, so I expect very few atoms to stick far down on the glass plate.[26]

In summary, I expect that atoms would leave the magnetic field in any of a range of deflections: a very few with large positive deflection, more with a small positive deflection, a lot with no deflection, some with a small negative deflection, and a very few with large negative deflection. This continuity of deflections reflects a continuity of magnetic moment projections.

[26]To be specific, this reasoning suggests that the number of atoms with moment tilted at angle θ relative to the z direction is proportional to $\sin\theta$, where θ ranges from 0° to 180°. You might want to prove this to yourself, but we'll never use this result so don't feel compelled.

In fact, however, this is not what happens at all! The projection μ_z does not take on a continuous range of values. Instead, it is quantized and takes on only two values, one positive and one negative. Those two values are called $\mu_z = \pm\mu_B$ where μ_B, the so-called "Bohr magneton", has the measured value of

$$\mu_B = 9.274\,010\,078 \times 10^{-24} \text{ J/T}, \tag{1.33}$$

with an uncertainty of 3 in the last decimal digit.

Distribution of μ_z

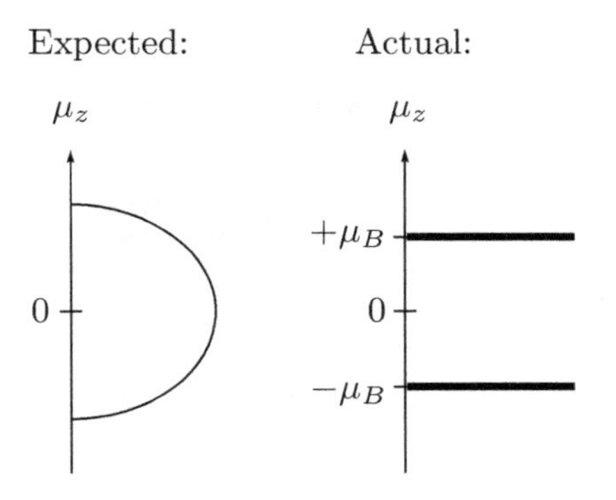

The Stern-Gerlach experiment was initially performed with silver atoms but has been repeated with many other types of atoms. When nitrogen is used, the projection μ_z takes on one of the four quantized values of $+3\mu_B$, $+\mu_B$, $-\mu_B$, or $-3\mu_B$. When sulfur is used, it takes on one of the five quantized values of $+4\mu_B$, $+2\mu_B$, 0, $-2\mu_B$, and $-4\mu_B$. For no atom do the values of μ_z take on the broad continuum of my classical expectation. For all atoms, the projection μ_z is quantized.

Problems

1.16 **Force on a classical magnetic moment**

The force on a classical magnetic moment is most easily calculated using "magnetic charge fiction": Consider the magnetic moment to consist of two "magnetic charges" of magnitude $+m$ and $-m$, separated by the position vector $\vec{d}$ running from $-m$ to $+m$. The magnetic moment is then $\vec{\mu} = m\vec{d}$.

a. Use B_+ for the magnitude of the magnetic field at $+m$, and B_- for the magnitude of the magnetic field at $-m$. Show that the net force on the magnetic moment is in the z direction with magnitude $mB_+ - mB_-$.
b. Use d_z for the z-component of the vector $\vec{d}$. Show that to high accuracy

$$B_+ = B_- + \frac{\partial B}{\partial z} d_z.$$

Surely, for distances of atomic scale, this accuracy is more than adequate.

c. Derive expression (1.32) for the force on a magnetic moment.

1.17 **Questions** (recommended problem)

Answering questions is an important scientific skill and, like any skill, it is sharpened through practice. This book gives you plenty of opportunities to develop that skill. Asking questions is another important scientific skill.[27] To hone that skill, write down a list of questions you have about quantum mechanics at this point. Be brief and pointed: you will not be graded for number or for verbosity. In future problems, I will ask you to add to your list.

⟦For example, one of my questions would be: "In ocean waves, the water is doing the 'waving'. In sound waves, it's the air. In light waves, it's the abstract electromagnetic field. What is 'waving' in a de Broglie wave?"⟧

[27] "The important thing is not to stop questioning," said Einstein. "Never lose a holy curiosity." [Interview by William Miller, "Death of a Genius", *Life* magazine, volume **38**, number 18 (2 May 1955) pages 61–64 on page 64.]

Chapter 2

What Is Quantum Mechanics About?

The story of Planck's discovery of the quantization of light energy is a fascinating one, but it's a difficult and elaborate story because it involves not just quantization, but also thermal equilibrium and electromagnetic radiation. The story of the discovery of atomic energy quantization is just as fascinating, but again fraught with intricacies. In an effort to remove the extraneous and dive deep to the heart of the matter, we focus on the measurement of the magnetic moment of a silver atom. We will, to the extent possible, do a quantum-mechanical treatment of an atom's magnetic moment while maintaining a classical treatment of all other aspects — such as its energy and momentum and position. (In chapter 4, "The Quantum Mechanics of Position", we take up a quantum-mechanical treatment of position and energy.)

The previous chapter attempted to apply classical pictures to atomic entities — electrons pictured as small, hard marbles; magnetic moments pictured as classical pointing arrows — and it found that the results were untenable. So this chapter will not impose the classical pictures in our minds onto nature. Instead we will perform experiments and let the atoms themselves tell us how they behave.

2.1 Quantization

2.1.1 *The conundrum of projections*

The Stern-Gerlach result — that μ_z is quantized rather than continuous — is counterintuitive and unexpected, but we can live with the counterintuitive and unexpected. It happens all the time in politics.

However, this fact of quantization appears to result in a logical contradiction, because there are many possible axes upon which the magnetic moment can be projected. The figures below make it clear that it is impossible for *any* vector to have a projection of either $\pm\mu_B$ on *all* axes!

Because if the projection of $\vec{\mu}$ on the z axis is $+\mu_B$ ...

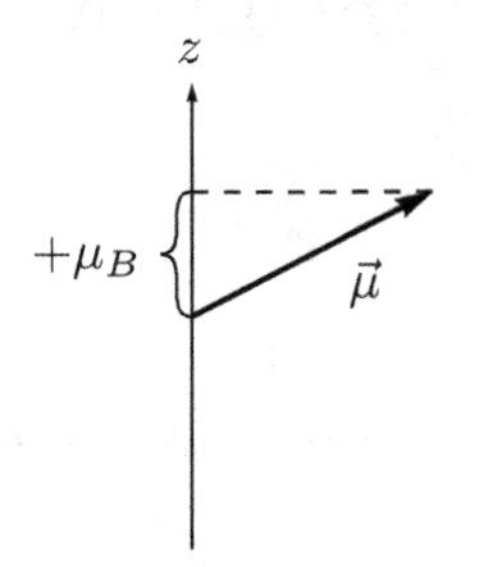

...then the projection of $\vec{\mu}$ on this second axis must be more than $+\mu_B$ ...

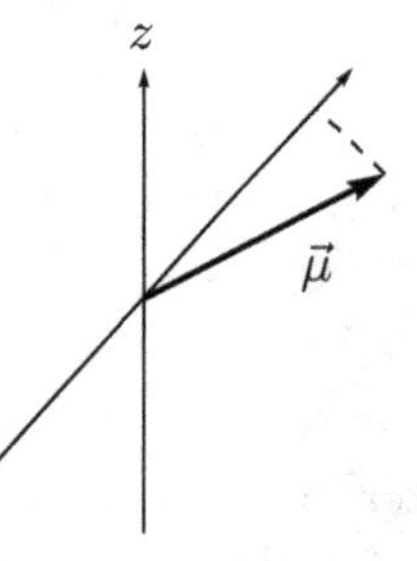

...while the projection of $\vec{\mu}$ on this third axis must be less than $+\mu_B$.

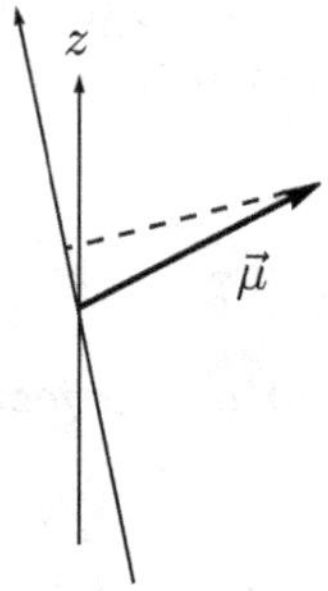

Whenever we measure the magnetic moment, projected onto any axis, the result is either $+\mu_B$ or $-\mu_B$. Yet is it impossible for the projection of any classical arrow on *all* axes to be either $+\mu_B$ or $-\mu_B$! This seeming contradiction is called "the conundrum of projections". We can live with the counterintuitive, the unexpected, the strange, but we cannot live with a logical contradiction. How can we resolve it?

The resolution comes not from meditating on the question, but from experimenting about it. Let us actually measure the projection on one axis, and then on a second. To do this easily, we modify the Stern-Gerlach apparatus and package it into a box called a "Stern-Gerlach analyzer". This box consists of a Stern-Gerlach apparatus followed by "pipes" that channel the outgoing atoms into horizontal paths.[1] This chapter treats only silver atoms, so we use analyzers with two exit ports.

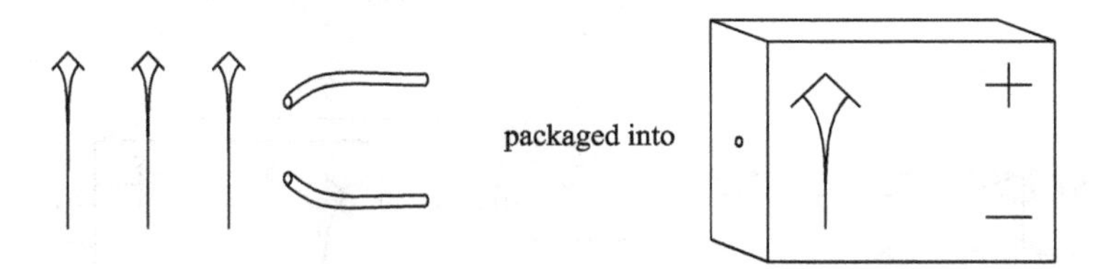

An atom enters a vertical analyzer through the single hole on the left. If it exits through the upper hole on the right (the "+ port") then the outgoing atom has $\mu_z = +\mu_B$. If it exits through the lower hole on the right (the "− port") then the outgoing atom has $\mu_z = -\mu_B$.

[1] In general, the "pipes" will manipulate the atoms through electromagnetic fields, not through touching. One way way to make such "pipes" is to insert a second Stern-Gerlach apparatus, oriented upside-down relative to the first. The atoms with $\mu_z = +\mu_B$, which had experienced an upward force in the first half, will experience an equal downward force in the second half, and the net impulse delivered will be zero. But whatever their manner of construction, the pipes must *not* change the magnetic moment of an atom passing through them.

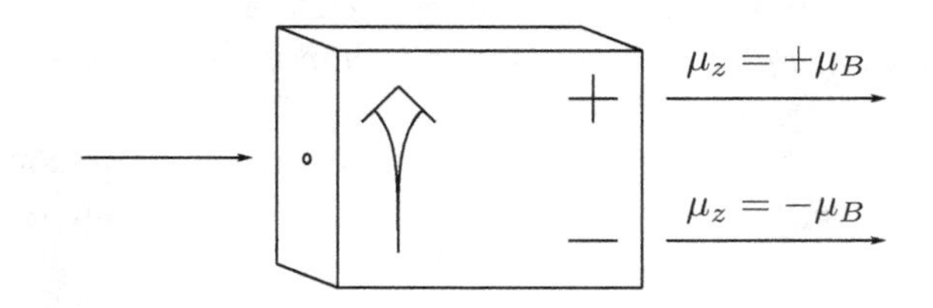

2.1.2 *Two vertical analyzers*

In order to check the operation of our analyzers, we do preliminary experiments. Atoms are fed into a vertical analyzer. Any atom exiting from the + port is then channeled into a second vertical analyzer. That atom exits from the + port of the second analyzer. This makes sense: the atom had $\mu_z = +\mu_B$ when exiting the first analyzer, and the second analyzer confirms that it has $\mu_z = +\mu_B$.

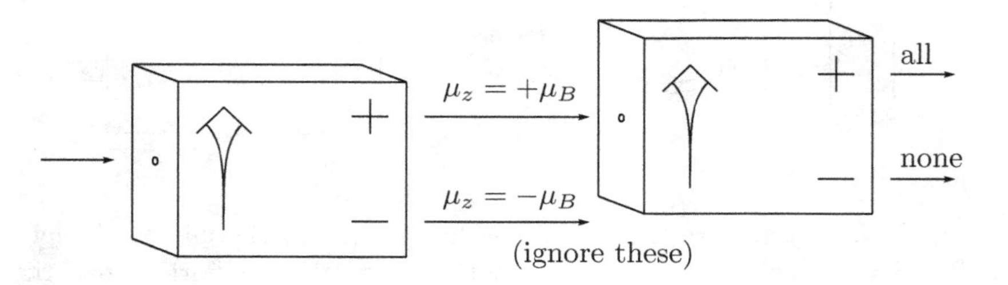

Furthermore, if an atom exiting from the − port of the first analyzer is channeled into a second vertical analyzer, then that atom exits from the − port of the second analyzer.

2.1.3 *One vertical and one upside-down analyzer*

Atoms are fed into a vertical analyzer. Any atom exiting from the + port is then channeled into a second analyzer, but this analyzer is oriented *upside-down.* What happens? If the projection on an upward-pointing axis is $+\mu_B$ (that is, $\mu_z = +\mu_B$), then the projection on a downward-pointing axis is $-\mu_B$ (we write this as $\mu_{(-z)} = -\mu_B$). So I expect that these atoms will emerge from the − port of the second analyzer (which happens to be the higher port). And this is exactly what happens.

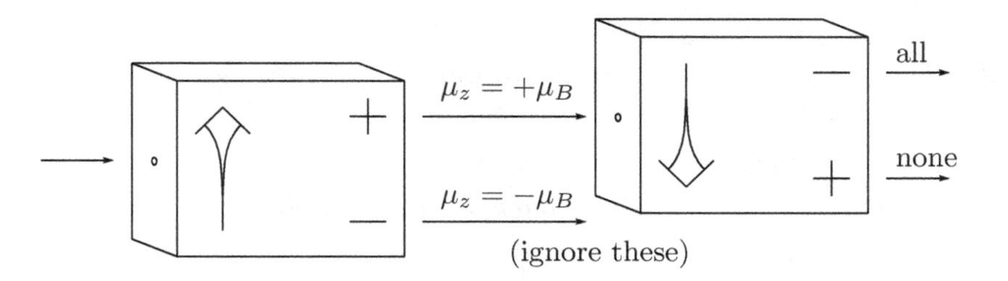

Similarly, if an atom exiting from the − port of the first analyzer is channeled into an upside-down analyzer, then that atom emerges from the + port of the second analyzer.

2.1.4 *One vertical and one horizontal analyzer*

Atoms are fed into a vertical analyzer. Any atom exiting from the + port is then channeled into a second analyzer, but this analyzer is oriented *horizontally.* The second analyzer doesn't measure the projection μ_z, it measures the projection μ_x. What happens in this case? Experiment shows that the atoms emerge randomly: half from the + port, half from the − port.

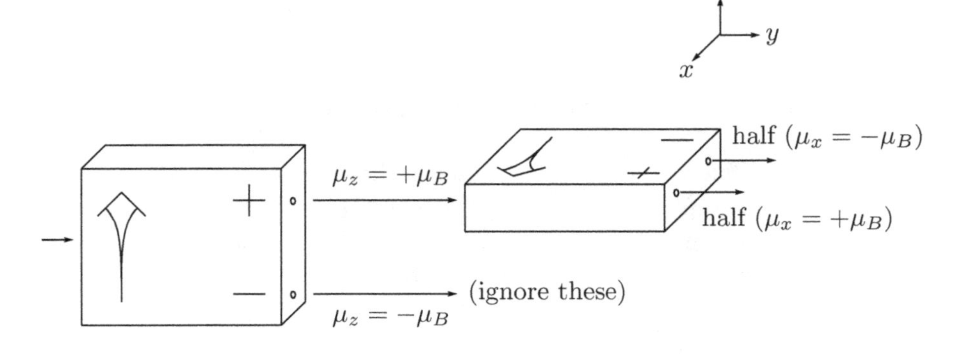

This makes some sort of sense. If a classical magnetic moment were vertically oriented, it would have $\mu_x = 0$, and such a classical moment would go straight through a horizontal Stern-Gerlach analyzer. We've seen that atomic magnetic moments *never* go straight through. If you "want" to go straight but are forced to turn either left or right, the best you can do is turn left half the time and right half the time. (Don't take this paragraph literally... atoms have no personalities and they don't "want" anything. But it is a useful mnemonic.)

2.1.5 *One vertical and one backwards horizontal analyzer*

Perform the same experiment as above (section 2.1.4), except insert the horizontal analyzer in the opposite sense, so that it measures the projection on the negative x axis rather than the positive x axis. Again, half the atoms emerge from the + port, and half emerge from the − port.

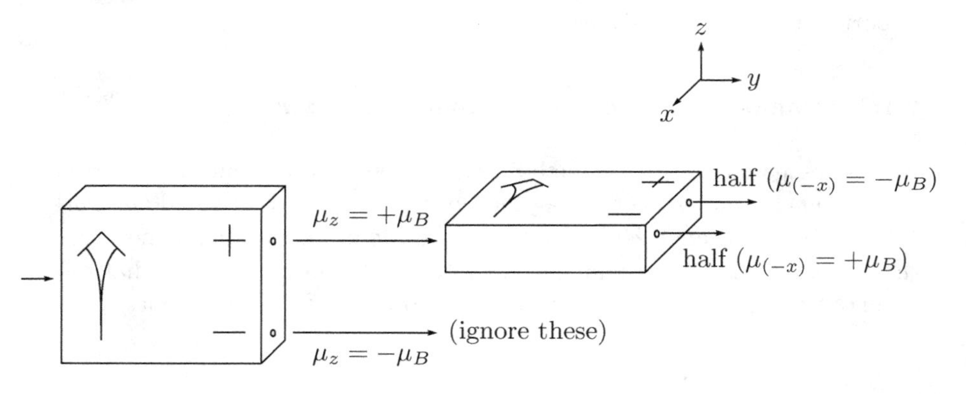

2.1.6 *One horizontal and one vertical analyzer*

A $+x$ analyzer followed by a $+z$ analyzer is the same apparatus as above (section 2.1.5), except that both analyzers are rotated as a unit by 90° about the y axis. So of course it has the same result: half the atoms emerge from the + port, and half emerge from the − port.

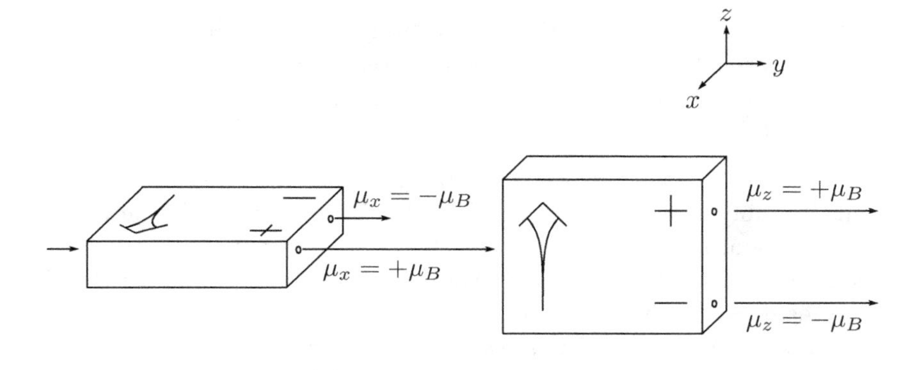

2.1.7 *Three analyzers*

Atoms are fed into a vertical analyzer. Any atom exiting from the + port is then channeled into a horizontal analyzer. Half of these atoms exit from the + port of the horizontal analyzer (see section 2.1.4), and *these* atoms are channeled into a third analyzer, oriented vertically. What happens at the third analyzer?

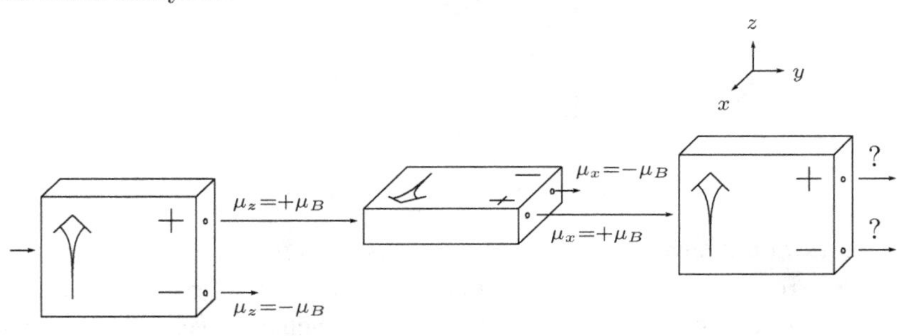

There are two ways to think of this: (I) When the atom emerged from the + port of the first analyzer, it was determined to have $\mu_z = +\mu_B$. When that same atom emerged from the + port of the second analyzer, it was determined to have $\mu_x = +\mu_B$. Now we know two projections of the magnetic moment. When it enters the third analyzer, it still has $\mu_z = +\mu_B$, so it will emerge from the + port. (II) The last two analyzers in this sequence are a horizontal analyzer followed by a vertical analyzer, and from section 2.1.6 we know what happens in this case: a 50/50 split. That will happen in this case, too.

So, analysis (I) predicts that all the atoms entering the third analyzer will exit through the + port and none through the − port. Analysis (II) predicts that half the atoms will exit through the + port and half through the − port.

Experiment shows that analysis (II) gives the correct result. But what could possibly be wrong with analysis (I)? Let's go through line by line: "When the atom emerged from the + port of the first analyzer, it was determined to have $\mu_z = +\mu_B$." Nothing wrong here — this is what an analyzer does. "When that same atom emerged from the + port of the second analyzer, it was determined to have $\mu_x = +\mu_B$." Ditto. "Now we know two projections of the magnetic moment." This has got to be the problem. To underscore that problem, look at the figure below.

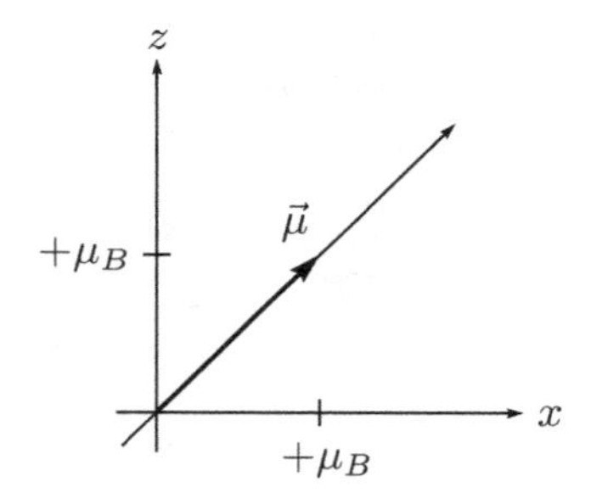

If an atom *did* have both $\mu_z = +\mu_B$ *and* $\mu_x = +\mu_B$, then the projection on an axis rotated 45° from the vertical would be $\mu_{45^\circ} = +\sqrt{2}\,\mu_B$. But the Stern-Gerlach experiment assures us that whenever μ_{45° is measured, the result is either $+\mu_B$ or $-\mu_B$, and never $+\sqrt{2}\,\mu_B$. In summary, *it is not possible for a moment to have a projection on both the z axis and on the x axis.* Passing to the fourth sentence of analysis (I) — "When the atom enters the third analyzer, it still has $\mu_z = +\mu_B$, so it will emerge from the + port" — we immediately see the problem. The atom emerging from the + port of the second analyzer does *not* have $\mu_z = +\mu_B$ — it doesn't have a projection on the z axis at all.

Because it's easy to fall into misconceptions, let me emphasize what I'm saying and what I'm not saying:

> I'm saying that if an atom has a value for μ_x, then it doesn't have a value for μ_z.
> I'm not saying that the atom has a value for μ_z but no one knows what it is.
> I'm not saying that the atom has a value for μ_z but that value is changing rapidly.
> I'm not saying that the atom has a value for μ_z but that value is changing unpredictably.
> I'm not saying that a random half of such atoms have the value $\mu_z = +\mu_B$ and the other half have the value $\mu_z = -\mu_B$.
> I'm not saying that the atom has a value for μ_z which will be disturbed upon measurement.

The atom with a value for μ_x does not have a value for μ_z in the same way that *love does not have a color.*

This is a new phenomenon, and it deserves a new name. That name is "indeterminacy". This is perhaps not the best name, because it might suggest, incorrectly, that an atom with a value for μ_x has a value for μ_z and we merely haven't yet determined what that value is. The English language was invented by people who didn't understand quantum mechanics, so it is not surprising that there are no perfectly appropriate names for quantum mechanical phenomena. This is a defect in our language, not a defect in quantum mechanics or in our understanding of quantum mechanics, and it is certainly not a defect in nature.[2]

How can a vector have a projection on one axis but not on another? It is the job of the rest of this book to answer that question, but one thing is clear already: The visualization of an atomic magnetic moment as a classical arrow must be wrong.

[2]In exactly the same manner, the name "orange" applies to light within the wavelength range 590–620 nm and the name"red" applies to light within the wavelength range 620–740 nm, but the English language has no word to distinguish the wavelength range 1590–1620 nm from the wavelength range 1620–1740 nm. This is not because optical light is "better" or "more deserving" than infrared light. It is due merely to the accident that our eyes detect optical light but not infrared light.

2.1.8 *The upshot*

We escape from the conundrum of projections through probability. If an atom has $\mu_z = +\mu_B$, and if the projection on some other axis is measured, then the result cannot be predicted with certainty: we instead give probabilities for the various results. If the second analyzer is rotated by angle θ relative to the vertical, the probability of emerging from the + port of the second analyzer is called $P_+(\theta)$.

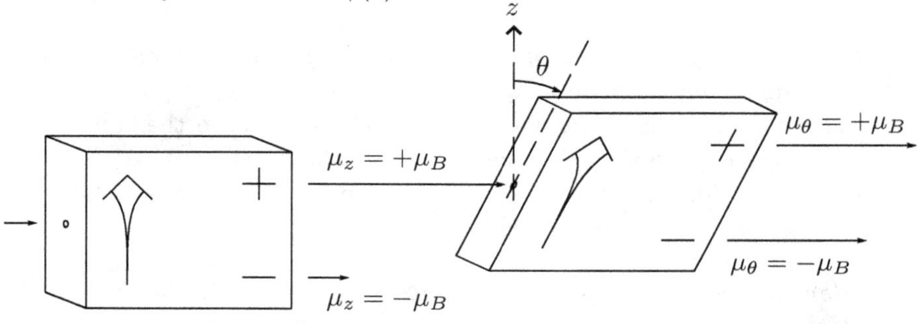

We already know some special values: from section 2.1.2, $P_+(0°) = 1$; from section 2.1.4, $P_+(90°) = \frac{1}{2}$; from section 2.1.3, $P_+(180°) = 0$; from section 2.1.5, $P_+(270°) = \frac{1}{2}$; from section 2.1.2, $P_+(360°) = 1$. It is not hard to guess the curve that interpolates between these values:

$$P_+(\theta) = \cos^2(\theta/2), \tag{2.1}$$

and experiment confirms this guess.

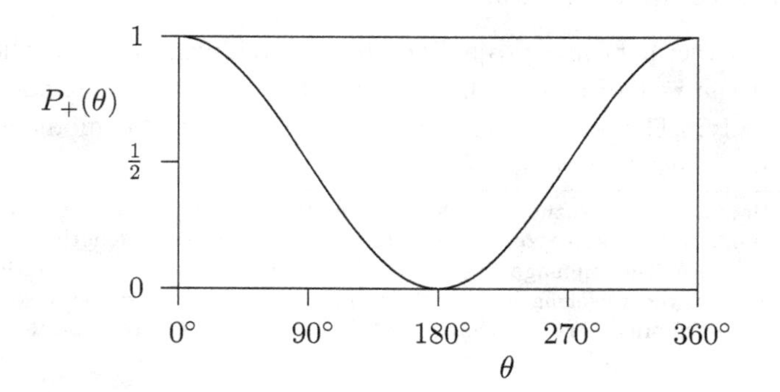

Problems

2.1 **Exit probabilities** (essential problem)

a. An analyzer is tilted from the vertical by angle α. An atom leaving its + port is channeled into a vertical analyzer. What is the probability that this atom emerges from the + port? The − port? (*Clue:* Use the "rotate as a unit" concept introduced in section 2.1.6.)

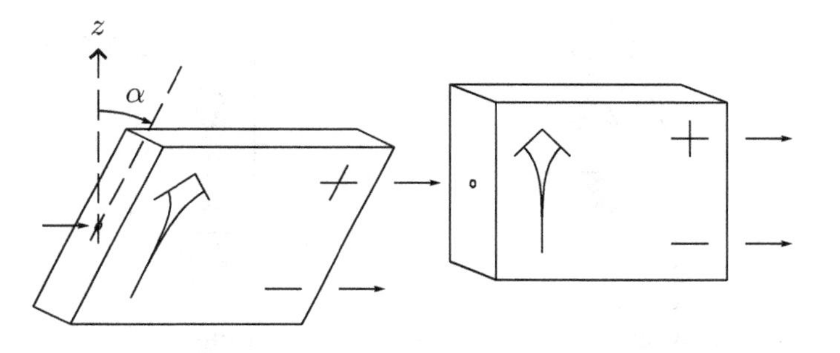

b. An atom exiting the − port of a vertical analyzer behaves exactly like one exiting the + port of an upside-down analyzer (see section 2.1.3). Such an atom is channeled into an analyzer tilted from the vertical by angle β. What is the probability that this atom emerges from the + port? The − port?

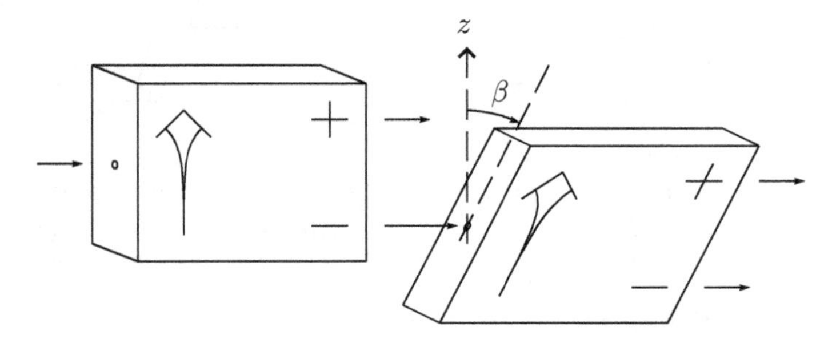

(Problem continues on next page.)

c. An analyzer is tilted from the vertical by angle γ. An atom leaving its $-$ port is channeled into a vertical analyzer. What is the probability that this atom emerges from the $+$ port? The $-$ port?

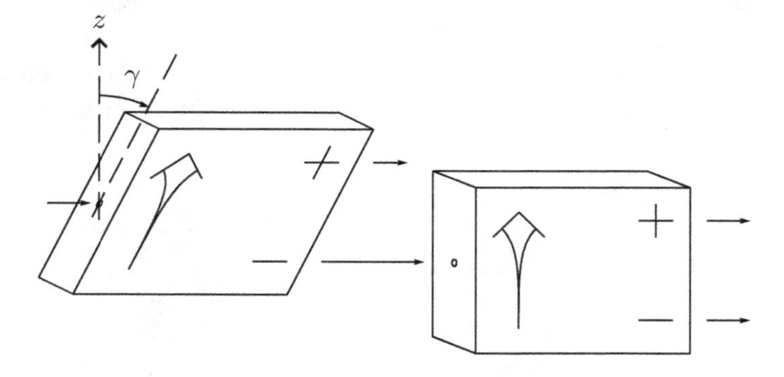

2.2 Multiple analyzers

An atom with $\mu_z = +\mu_B$ is channeled through the following line of three Stern-Gerlach analyzers.

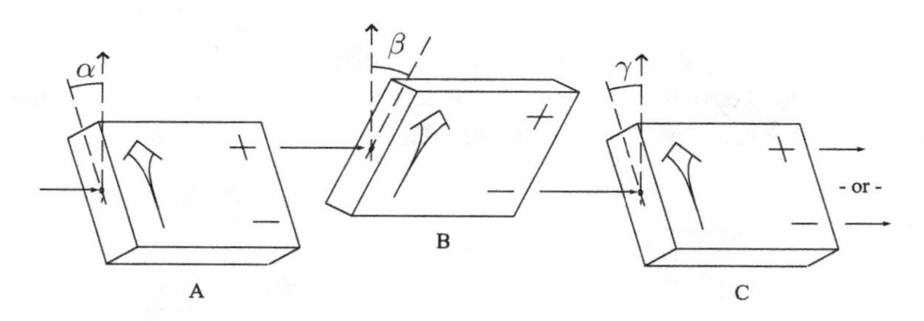

Find the probability that it emerges from (a) the $-$ port of analyzer A; (b) the $+$ port of analyzer B; (c) the $+$ port of analyzer C; (d) the $-$ port of analyzer C.

2.3 Properties of the $P_+(\theta)$ function

a. An atom exits the $+$ port of a vertical analyzer; that is, it has $\mu_z = +\mu_B$. Argue that the probability of this atom exiting from the $-$ port of a θ analyzer is the same as the probability of it exiting from the $+$ port of a $(180^\circ - \theta)$ analyzer.

b. Conclude that the $P_+(\theta)$ function introduced in section 2.1.8 must satisfy

$$P_+(\theta) + P_+(180^\circ - \theta) = 1.$$

c. Does the experimental result (2.1) satisfy this condition?

2.2 Interference

There are more quantum mechanical phenomena to uncover. To support our exploration, we build a new experimental device called the "analyzer loop".[3] This is nothing but a Stern-Gerlach analyzer followed by "piping" that channels the two exit paths together again.[4]

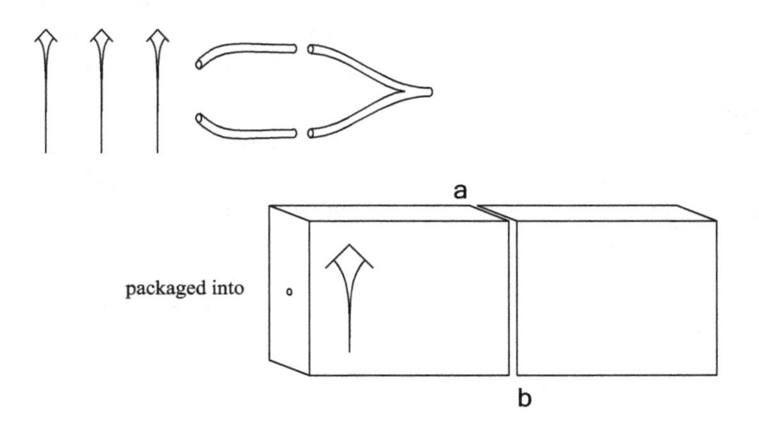

The device must be constructed to high precision, so that there can be no way to distinguish whether the atom passed through by way of the top path or the bottom path. For example, the two paths must have the same length: If the top path were longer, then an atom going through via the top path would take more time, and hence there *would* be a way to tell which way the atom passed through the analyzer loop.

In fact, the analyzer loop is constructed so precisely that it doesn't change the character of the atom passing through it. If the atom enters

[3]We build it in our minds. The experiments described in this section have never been performed exactly as described here, although researchers are getting close. [See Shimon Machluf, Yonathan Japha, and Ron Folman, "Coherent Stern–Gerlach momentum splitting on an atom chip" *Nature Communications* **4** (9 September 2013) 2424.] We know the results that would come from these experiments because conceptually parallel (but more complex!) experiments *have* been performed on photons, neutrons, atoms, and molecules. (See page 33.)

[4]If you followed the footnote on page 49, you will recall that these "pipes" manipulate atoms through electromagnetic fields, not through touching. One way to make them would be to insert two more Stern-Gerlach apparatuses, the first one upside-down and the second one rightside-up relative to the initial apparatus. But whatever the manner of their construction, the pipes must *not* change the magnetic moment of an atom passing through them.

with $\mu_z = +\mu_B$, it exits with $\mu_z = +\mu_B$. If it enters with $\mu_x = -\mu_B$, it exits with $\mu_x = -\mu_B$. If it enters with $\mu_{17°} = -\mu_B$, it exits with $\mu_{17°} = -\mu_B$. It is hard to see why anyone would want to build such a device, because they're expensive (due to the precision demands), and they do absolutely nothing!

Once you made one, however, you could convert it into something useful. For example, you could insert a piece of metal blocking path a. In that case, all the atoms exiting would have taken path b, so (if the analyzer loop were oriented vertically) all would emerge with $\mu_z = -\mu_B$.

Using the analyzer loop, we set up the following apparatus: First, channel atoms with $\mu_z = +\mu_B$ into a horizontal analyzer loop.[5] Then, channel the atoms emerging from that analyzer loop into a vertical analyzer. Ignore atoms emerging from the + port of the vertical analyzer and look for atoms emerging from the − port.

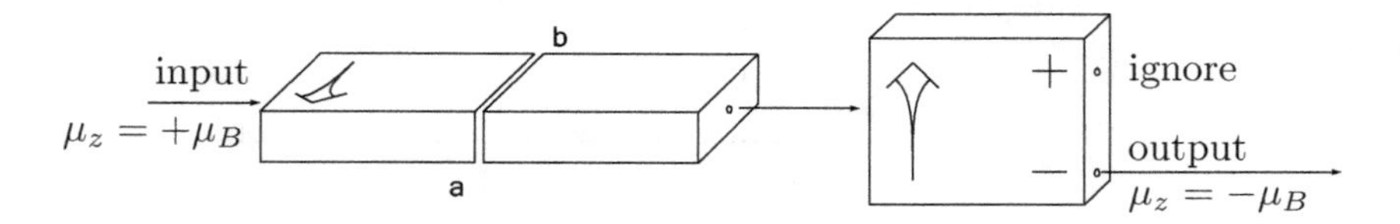

We execute three experiments with this set-up: first we pass atoms through when path a is blocked, then when path b is blocked, finally when neither path is blocked.

2.2.1 *Path* a *blocked*

(1) Atoms enter the analyzer loop with $\mu_z = +\mu_B$.
(2) Half of them attempt path a, and end up impaled on the blockage.
(3) The other half take path b, and emerge from the analyzer loop with $\mu_x = -\mu_B$.
(4) Those atoms then enter the vertical analyzer. Similar to the result of section 2.1.6, half of these atoms emerge from the + port and are ignored. Half of them emerge from the − port and are counted.
(5) The overall probability of passing through the set-up is $\frac{1}{2} \times \frac{1}{2} = \frac{1}{4}$.

If you perform this experiment, you will find that this analysis is correct and that these results are indeed obtained.

[5]To make sure that all of these atoms have $\mu_z = +\mu_B$, they are harvested from the + port of a vertical analyzer.

2.2.2 *Path* b *blocked*

(1) Atoms enter the analyzer loop with $\mu_z = +\mu_B$.
(2) Half of them attempt path b, and end up impaled on the blockage.
(3) The other half take path a, and emerge from the analyzer loop with $\mu_x = +\mu_B$.
(4) Those atoms then enter the vertical analyzer. Exactly as in section 2.1.6, half of these atoms emerge from the + port and are ignored. Half of them emerge from the − port and are counted.
(5) The overall probability of passing through the set-up is $\frac{1}{2} \times \frac{1}{2} = \frac{1}{4}$.

Once again, experiment confirms these results.

2.2.3 *Neither path blocked*

Here, I have not just one, but two ways to analyze the experiment:

Analysis I:

(1) An atom passes through the set-up either via path b or via path a.
(2) From section 2.2.1, the probability of passing through via path b is $\frac{1}{4}$.
(3) From section 2.2.2, the probability of passing through via path a is $\frac{1}{4}$.
(4) Thus the probability of passing through the entire set-up is $\frac{1}{4} + \frac{1}{4} = \frac{1}{2}$.

Analysis II:

(1) Because "the analyzer loop is constructed so precisely that it doesn't change the character of the atom passing through it", the atom emerges from the analyzer loop with $\mu_z = +\mu_B$.
(2) When such atoms enter the vertical analyzer, all of them emerge through the + port. (See section 2.1.2.)
(3) Thus the probability of passing through the entire set-up is zero.

These two analyses cannot both be correct. Experiment confirms the result of analysis II, but what could possibly be wrong with analysis I? Item (2) is already confirmed through the experiment of section 2.2.1, item (3) is already confirmed through the experiment of section 2.2.2, and don't tell me that I made a mistake in the arithmetic of item (4). The only thing left is item (1): "An atom passes through the set-up either via path b or via path a." This simple, appealing, common-sense statement must be *wrong*!

Just a moment ago, the analyzer loop seemed like a waste of money and skill. But in fact, a horizontal analyzer loop is an extremely clever way of correlating the path through the analyzer loop with the value of μ_x: If the atom has $\mu_x = +\mu_B$, then it takes path a. If the atom has $\mu_x = -\mu_B$, then it takes path b. If the atom has $\mu_z = +\mu_B$, then it *doesn't have a value of* μ_x and hence it *doesn't take a path.*

Notice again what I'm saying: I'm not saying the atom takes one path or the other but we don't know which. I'm not saying the atom breaks into two pieces and each half traverses its own path. I'm saying the atom doesn't take a path. The $\mu_z = +\mu_B$ atoms within the horizontal analyzer loop *do not have a position* in the same sense that *love does not have a color.* If you think of an atom as a smaller, harder version of a classical marble, then you're visualizing the atom incorrectly.

Once again, our experiments have uncovered a phenomenon that doesn't happen in daily life, so there is no word for it in conventional language.[6] Sometimes people say that "the atom takes both paths", but that phrase does not really get to the heart of the new phenomenon. I have asked students to invent a new word to represent this new phenomenon, and my favorite of their many suggestions is "ambivate" — a combination of ambulate and ambivalent — as in "an atom with $\mu_z = +\mu_B$ ambivates through both paths of a horizontal analyzer loop". While this is a great word, it hasn't caught on. The conventional name for this phenomenon is "quantal interference".

The name "quantal interference" comes from a (far-fetched) analogy with interference in wave optics. Recall that in the two-slit interference of light, there are some observation points that have a light intensity if light passes through slit a alone, and the same intensity if light passes through slit b alone, but zero intensity if light passes through both slits. This is called "destructive interference". There are other observation points that have a light intensity if the light passes through slit a alone, and the same intensity if light passes through slit b alone, but four times that intensity if

[6]In exactly the same way, there was no need for the word "latitude" or the word "longitude" when it was thought that the Earth was flat. The discovery of the near-spherical character of the Earth forced our forebears to invent new words to represent these new concepts. Words do not determine reality; instead reality determines which words are worth inventing.

light passes through both slits. This is called "constructive interference". But in fact the word "interference" is a poor name for this phenomenon as well. It's adapted from a football term, and football players never (or at least never intentionally) run "constructive interference".

One last word about language: The device that I've called the "analyzer loop" is more conventionally called an "interferometer". I didn't use that name at first because that would have given away the ending.

Back on page 47 I said that, to avoid unnecessary distraction, this chapter would "to the extent possible, do a quantum-mechanical treatment of an atom's magnetic moment while maintaining a classical treatment of all other aspects — such as its energy and momentum and position". You can see now why I put in that qualifier "to the extent possible": we have found that within an interferometer, a quantum-mechanical treatment of magnetic moment demands a quantum-mechanical treatment of position as well.

2.2.4 *Sample Problem: Constructive interference*

Consider the same set-up as on page 60, but now ignore atoms leaving the $-$ port of the vertical analyzer and consider as output atoms leaving the $+$ port. What is the probability of passing through the set-up when path a is blocked? When path b is blocked? When neither path is blocked?

Solution: $\frac{1}{4}$; $\frac{1}{4}$; 1. Because $\frac{1}{4}+\frac{1}{4}<1$, this is an example of constructive interference.

2.2.5 *Sample Problem: Two analyzer loops*

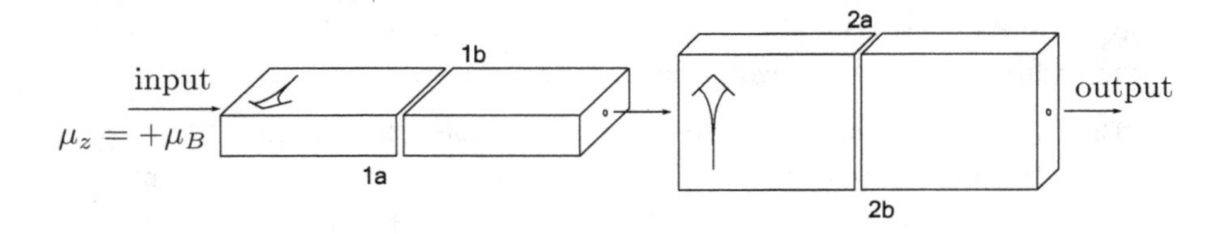

Atoms with $\mu_z = +\mu_B$ are channeled through a horizontal analyzer loop (number 1), then a vertical analyzer loop (number 2). If all paths are open, 100% of the incoming atoms exit from the output. What percentage of the incoming atoms leave from the output if the following paths are blocked?

(a)	2a	(d)	1b
(b)	2b	(e)	1b and 2a
(c)	1a	(f)	1a and 2b

Solution: Only two principles are needed to solve this problem: First, an atom leaving an unblocked analyzer loop leaves in the same condition it had when it entered. Second, an atom leaving an analyzer loop with one path blocked leaves in the condition specified by the path that it took, regardless of the condition it had when it entered. Use of these principles gives the solution in the table on the next page. Notice that in changing from situation (a) to situation (e), you add blockage, yet you increase the output!

paths blocked	input condition	path taken through # 1	intermediate condition	path taken through # 2	output condition	probability of input $\to$ output
none	$\mu_z = +\mu_B$	"both"	$\mu_z = +\mu_B$	a	$\mu_z = +\mu_B$	100%
2a	$\mu_z = +\mu_B$	"both"	$\mu_z = +\mu_B$	100% blocked at a	none	0%
2b	$\mu_z = +\mu_B$	"both"	$\mu_z = +\mu_B$	a	$\mu_z = +\mu_B$	100%
1a	$\mu_z = +\mu_B$	50% blocked at a 50% pass through b	$\mu_x = -\mu_B$	"both"	$\mu_x = -\mu_B$	50%
1b	$\mu_z = +\mu_B$	50% pass through a 50% blocked at b	$\mu_x = +\mu_B$	"both"	$\mu_x = +\mu_B$	50%
1b and 2a	$\mu_z = +\mu_B$	50% pass through a 50% blocked at b	$\mu_x = +\mu_B$	25% blocked at a 25% pass through b	$\mu_z = -\mu_B$	25%
1a and 2b	$\mu_z = +\mu_B$	50% blocked at a 50% pass through b	$\mu_x = -\mu_B$	25% pass through a 25% blocked at b	$\mu_z = +\mu_B$	25%

Problems

2.4 **Tilted analyzer loop** (recommended problem)

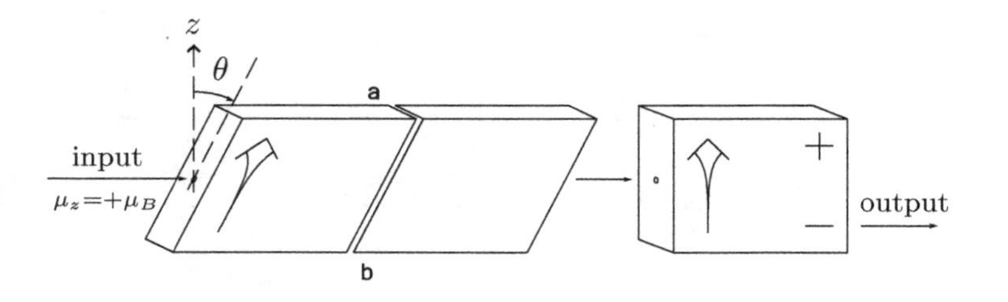

An atom with $\mu_z = +\mu_B$ enters the analyzer loop (interferometer) shown above, tilted at angle θ to the vertical. The outgoing atom enters a z-analyzer, and whatever comes out the − port is considered output. What is the probability for passage from input to output when:

a. Paths a and b are both open?
b. Path b is blocked?
c. Path a is blocked?

2.5 **Find the flaw: Tilted analyzer loop**[7]
Five students — Aldo, Beth, Celine, Denzel, and Ellen — work the above problem. All find the same answer for part (a), namely zero, but for parts (b) and (c) they produce five different answers! Their candidate answers are:

	(b)	(c)
Aldo	$\cos^4(\theta/2)$	$\sin^4(\theta/2)$
Beth	$\frac{1}{4}\sin^2(\theta)$	$\frac{1}{4}\sin^2(\theta)$
Celine	$\frac{1}{4}\sin(\theta)$	$\frac{1}{4}\sin(\theta)$
Denzel	$\frac{1}{4}\sqrt{2}\sin(\theta/2)$	$\frac{1}{4}\sqrt{2}\sin(\theta/2)$
Ellen	$\frac{1}{2}\sin^2(\theta)$	$\frac{1}{2}\sin^2(\theta)$

Provide simple reasons showing that four of these candidates must be wrong.

[7]Background concerning "find the flaw" type problems is provided in sample problem 1.2.1 on page 26.

2.6 **Three analyzer loops** (recommended problem)
Atoms with $\mu_z = +\mu_B$ are channeled into a horizontal analyzer loop, followed by a vertical analyzer loop, followed by a horizontal analyzer loop.

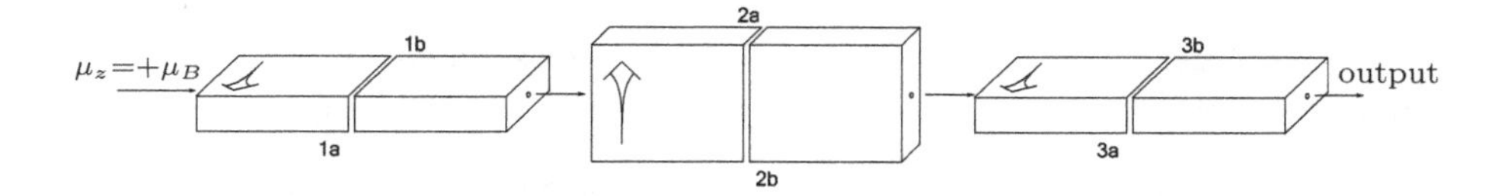

If all paths are open, 100% of the incoming atoms exit from the output. What percent of the incoming atoms leave from the output if the following paths are blocked?

(a)	3a	(d)	2b	(g)	1b and 3b
(b)	3b	(e)	1b	(h)	1b and 3a
(c)	2a	(f)	2a and 3b	(i)	1b and 3a and 2a

(Note that in going from situation (h) to situation (i) you get *more* output from *increased* blockage.)

2.3 Aharonov-Bohm effect

We have seen how to sort atoms using a Stern-Gerlach analyzer, made of a non-uniform magnetic field. But how do atoms behave in a uniform magnetic field? In general, this is an elaborate question, and the answer will depend on the initial condition of the atom's magnetic moment, on the magnitude of the field, and on the amount of time that the atom spends in the field. But for one special case the answer, determined experimentally, is easy. If an atom is exposed to uniform magnetic field $\vec{B}$ for exactly the right amount of time [which turns out to be a time of $h/(2\mu_B B)$], then the atom emerges with exactly the same magnetic condition it had initially: If it starts with $\mu_z = -\mu_B$, it ends with $\mu_z = -\mu_B$. If it starts with $\mu_x = +\mu_B$, it ends with $\mu_x = +\mu_B$. If it starts with $\mu_{29°} = +\mu_B$, it ends with $\mu_{29°} = +\mu_B$. Thus for atoms moving at a given speed, we can build a box containing a uniform magnetic field with just the right length so that any atom passing through it will spend just the right amount of time to emerge in the same condition it had when it entered. We call this box a "replicator".

If you play with one of these boxes you'll find that you can build any elaborate set-up of sources, detectors, blockages, and analyzers, and that inserting a replicator into any path will not affect the outcome of any experiment. But notice that this apparatus list does not include interferometers (our "analyzer loops")! Build the interference experiment of page 60. Do not block either path. Instead, slip a replicator into one of the two paths a or b — it doesn't matter which.

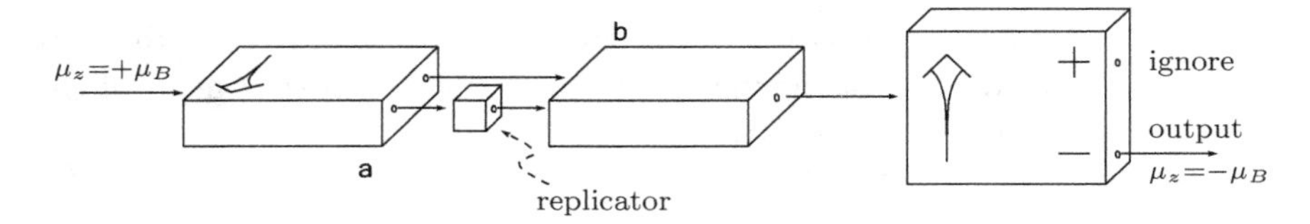

Without the replicator *no* atom emerges at output. But experiment shows that after inserting the replicator, *all* the atoms emerge at output.

How can this be? Didn't we just say of a replicator that "any atom passing through it will... emerge in the same condition it had when it entered"? Indeed we did, and indeed this is true. But an atom with $\mu_z = +\mu_B$ doesn't pass through path a or path b — it ambivates through both paths.

If the atom *did* take one path or the other, then the replicator *would* have no effect on the experimental results. The fact that it *does* have an effect is proof that the atom *doesn't* take one path or the other.

The fact[8] that one can perform this remarkable experiment was predicted theoretically (in a different context) by Walter Franz. He announced his result in Danzig (now Gdańsk, Poland) in May 1939, just months before the Nazi invasion of Poland, and his prediction was largely forgotten in the resulting chaos. The effect was rediscovered theoretically by Werner Ehrenberg and Raymond Siday in 1949, but they published their result under the opaque title of "The refractive index in electron optics and the principles of dynamics" and their prediction was also largely forgotten. The effect was rediscovered theoretically a third time by Yakir Aharonov and David Bohm in 1959, and this time it sparked enormous interest, both experimental and theoretical. The phenomenon is called today the "Aharonov-Bohm effect".

[8]See B.J. Hiley, "The early history of the Aharonov-Bohm effect" (17 April 2013) https://arxiv.org/abs/1304.4736.

Problem

2.7 **Bomb-testing interferometer**[9] (recommended problem)

The Acme Bomb Company sells a bomb triggered by the presence of silver, and claims that the trigger is so sensitive that the bomb explodes when its trigger absorbs even a single silver atom. You have heard similar extravagant claims from other manufacturers, so you're suspicious. You purchase a dozen bombs, then shoot individual silver atoms at each in turn. The first bomb tested explodes! The trigger worked as advertised, but now it's useless because it's blasted to pieces. The second bomb tested doesn't explode — the atom slips through a hole in the trigger. This confirms your suspicion that not *all* the triggers are as sensitive as claimed, so this bomb is useless to you as well. If you continue testing in this fashion, at the end all your good bombs will be blown up and you will be left with a stash of bad bombs.

So instead, you set up the test apparatus sketched here:

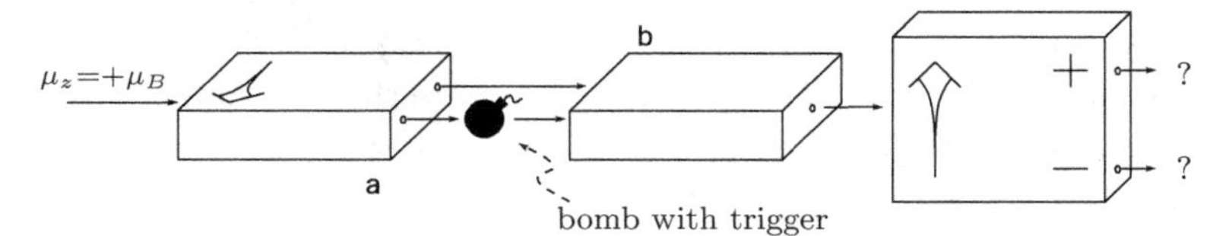

An atom with $\mu_z = +\mu_B$ enters the interferometer. If the bomb trigger has a hole, then the atom ambivates through both paths, arrives at the analyzer with $\mu_z = +\mu_B$, and exits the + port of the analyzer. But if the bomb trigger is good, then either (a) the atom takes path a and sets off the bomb, or else (b) the atom takes path b.

a. If the bomb trigger is good, what is the probability of option (a)? Of option (b)?

b. If option (b) happens, what kind of atom arrives at the analyzer? What is the probability of that atom exiting through the + port? The − port?

Conclusion: If the atom exits through the − port, then the bomb is good. If it exits through the + port then the bomb might be good or bad and further testing is required. But you *can* determine that the bomb trigger is good without blowing it up!

[9] Avshalom C. Elitzur and Lev Vaidman, "Quantum mechanical interaction-free measurements" *Foundations of Physics* **23** (July 1993) 987–997.

2.4 Light on the atoms

Our conclusion that, under some circumstances, the atom "does not have a position" is so dramatically counterintuitive that you might — no, you *should* — be tempted to test it experimentally. Set up the interference experiment on page 60, but instead of simply allowing atoms to pass through the interferometer, watch to see which path the atom takes through the set-up. To watch them, you need light. So set up the apparatus with lamps trained on the two paths a and b.

Send in one atom. There's a flash of light at path a.

Another atom. Flash of light at b.

Another atom. Flash at b again.

Then a, then a, then b.

You get the drift. Always the light appears at one path or the other. (In fact, the flashes come at random with probability $\frac{1}{2}$ for a flash at a and $\frac{1}{2}$ for a flash at b.) Never is there no flash. Never are there "two half flashes". The atom always has a position when passing through the interferometer. "So much", say the skeptics, "for this metaphysical nonsense about 'the atom takes both paths'."

But wait. Go back and look at the output of the vertical analyzer. When we ran the experiment with no light, the probability of coming out the $-$ port was 0. When we turn the lamps on, then the probability of coming out the $-$ port becomes $\frac{1}{2}$.

When the lamps are off, analysis II on page 61 is correct: the atoms ambivate through both paths, and the probability of exiting from the $-$ port is 0. When the lamps are on and a flash is seen at path a, then the atom *does* take path a, and now the analysis of section 2.2.2 on page 61 is correct: the probability of exiting from the $-$ port is $\frac{1}{2}$.

The process when the lamps are on is called "observation" or "measurement", and a lot of nonsense has come from the use of these two words. The important thing is whether the light is present or absent. Whether or not the flashes are "observed" by a person is irrelevant. To prove this to yourself, you may, instead of observing the flashes in person, record the flashes on video. If the lamps are on, the probability of exiting from the $-$ port is $\frac{1}{2}$. If the lamps are off, the probability of exiting from the $-$ port is 0. Now, after the experiment is performed, you may either destroy the video, or play it back to a human audience, or play it back to a feline audience. Surely, by this point it is too late to change the results at the exit port.

It's not just light. Any method you can dream up for determining the path taken will show that the atom takes just one path, but that method will also change the output probability from 0 to $\frac{1}{2}$. No person needs to actually read the results of this mechanism: as long as the mechanism is at work, as long as it is in principle possible to determine which path is taken, then one path is taken and no interference happens.

What happens if you train a lamp on path a but leave path b in the dark? In this case a flash means the atom has taken path a. No flash means the atom has taken path b. In both cases the probability of passage for the atom is $\frac{1}{2}$.

How can the atom taking path b "know" that the lamp at path a is turned on? The atom initially "sniffs out" *both* paths, like a fog creeping down two passageways. The atom that eventually does take path b in the dark started out attempting both paths, and that's how it "knows" the lamp at path a is on. This is called the "Renninger negative-result experiment".

It is not surprising that the presence or absence of light should affect an atom's motion: this happens even in classical mechanics. When an object absorbs or reflects light, that object experiences a force, so its motion is altered. For example, a baseball tossed upward in a gymnasium with the overhead lamps off attains a slightly greater height that an identical baseball experiencing an identical toss in the same gymnasium with the overhead lamps on, because the downward-directed light beams push the baseball downward. (This is the same "radiation pressure" that is responsible for the tails of comets. And of course, the effect occurs whenever the lamps are turned on: whether any person actually watches the illuminated baseball is irrelevant.) This effect is negligible for typical human-scale baseballs and tosses and lamps, but atoms are far smaller than baseballs and it is reasonable that the light should alter the motion of an atom more than it alters the motion of a baseball.

One last experiment: Look for the atoms with dim light. In this case, some of the atoms will pass through with a flash. But — because of the dimness — some atoms will pass through without any flash at all. For those atoms passing through with a flash, the probability for exiting the $-$ port is $\frac{1}{2}$. For those atoms passing through without a flash, the probability of exiting the $-$ port is 0.

2.5 Entanglement

I have claimed that an atom with $\mu_z = +\mu_B$ *doesn't have* a value of μ_x, and that when such an atom passes through a horizontal interferometer, it *doesn't have* a position. You might say to yourself, "These claims are so weird, so far from common sense, that I just can't accept them. I believe the atom *does* have a value of μ_x and *does* have a position, but something else very complicated is going on to make the atom *appear* to lack a μ_x and a position. I don't know what that complicated thing is, but just because I haven't yet thought it up yet doesn't mean that it doesn't exist."

If you think this, you're in good company: Einstein thought it too. This section introduces a new phenomenon of quantum mechanics, and shows that no local deterministic mechanism, no matter how complex or how fantastic, can give rise to all the results of quantum mechanics. Einstein was wrong.

2.5.1 *Flipping Stern-Gerlach analyzer*

A new piece of apparatus helps us uncover this new phenomenon of nature. Mount a Stern-Gerlach analyzer on a stand so that it can be oriented either vertically (0°), or tilted one-third of a circle clockwise (+120°), or tilted one-third of a circle counterclockwise (−120°). Call these three orientations V (for vertical), O (for out of the page), or I (for into the page). As an atom approaches the analyzer, select one of these three orientations at random, flip the analyzer to that orientation, and allow the atom to pass through as usual. As a new atom approaches, again select an orientation at random, flip the analyzer, and let the atom pass through. Repeat many times.

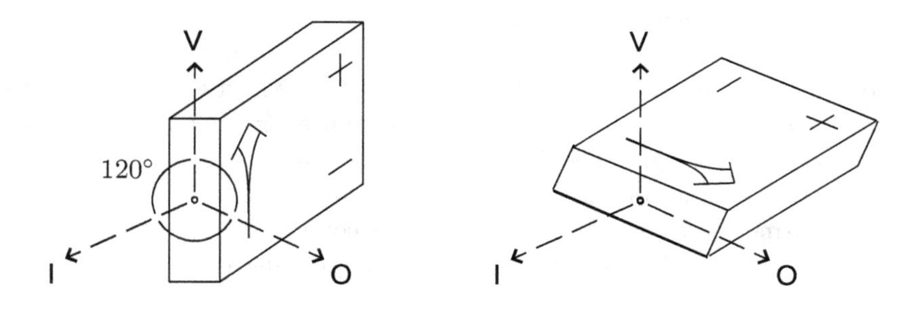

Flipping Stern-Gerlach analyzer. The arrows V*,* O*, and* I*, oriented* 120° *apart, all lie within the plane perpendicular to the atom's approach path.*

What happens if an atom with $\mu_z = +\mu_B$ enters a flipping analyzer? With probability $\frac{1}{3}$, the atom enters a vertical analyzer (orientation V), and in that case it exits the + port with probability 1. With probability $\frac{1}{3}$, the atom enters an out-of-the-page analyzer (orientation O), and in that case (see equation 2.1) it exits the + port with probability

$$\cos^2(120°/2) = \tfrac{1}{4}.$$

With probability $\frac{1}{3}$, the atom enters an into-the-page analyzer (orientation I), and in that case it exits the + port with probability $\frac{1}{4}$. Thus the overall probability of this atom exiting through the + port is

$$\tfrac{1}{3} \times 1 + \tfrac{1}{3} \times \tfrac{1}{4} + \tfrac{1}{3} \times \tfrac{1}{4} = \tfrac{1}{2}. \tag{2.2}$$

A similar analysis shows that if an atom with $\mu_z = -\mu_B$ enters the flipping analyzer, it exits the + port with probability $\frac{1}{2}$.

You could repeat the analysis for an atom entering with $\mu_{(+120°)} = +\mu_B$, but you don't need to. Because the three orientations are exactly one-third of a circle apart, rotational symmetry demands that an atom entering with $\mu_{(+120°)} = +\mu_B$ behaves exactly as an atom entering with $\mu_z = +\mu_B$.

In conclusion, an atom entering in any of the six conditions $\mu_z = +\mu_B$, $\mu_z = -\mu_B$, $\mu_{(+120°)} = +\mu_B$, $\mu_{(+120°)} = -\mu_B$, $\mu_{(-120°)} = +\mu_B$, or $\mu_{(-120°)} = -\mu_B$ will exit through the + port with probability $\frac{1}{2}$.

2.5.2 *EPR source of atom pairs*

Up to now, our atoms have come from an oven. For the next experiments we need a special source[10] that expels two atoms at once, one moving to the left and the other to the right. For the time being we call this an "EPR" source, which produces an atomic pair in an "EPR" condition. The letters come from the names of those who discovered this condition: Albert Einstein, Boris Podolsky, and Nathan Rosen. After investigating this condition we will develop a more descriptive name.

The following four experiments investigate the EPR condition:

(1) Each atom encounters a vertical Stern-Gerlach analyzer. The experimental result: *the two atoms exit through opposite ports.* To be precise: with probability $\frac{1}{2}$, the left atom exits + and the right atom exits −, and with probability $\frac{1}{2}$, the left atom exits − and the right atom exits +, but it never happens that both atoms exit + or that both atoms exit −.

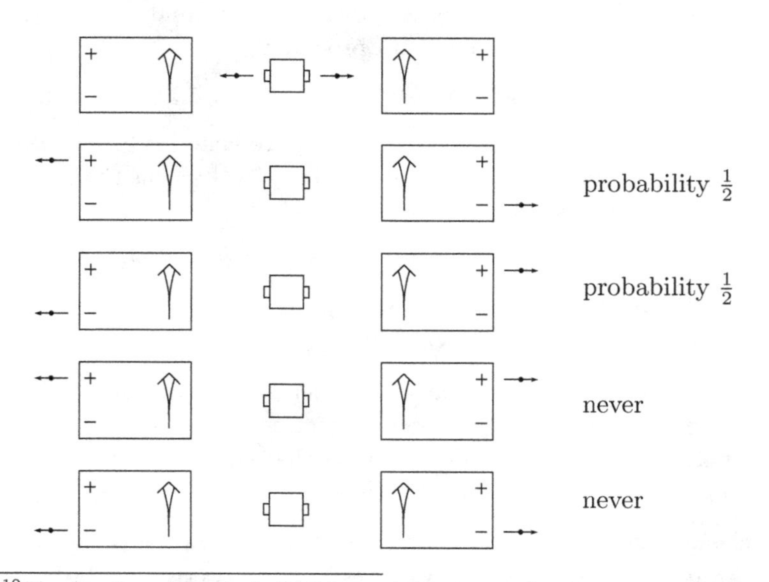

[10]The question of how to build this special source need not concern us at the moment: it is an experimental fact that such sources do exist. One way to make one would start with a diatomic molecule with zero magnetic moment. Cause the molecule to disintegrate and eject the two daughter atoms in opposite directions. Because the initial molecule had zero magnetic moment, the pair of daughter atoms will have the properties of magnetic moment described. In fact, it's easier to build a source, not for a pair of atoms, but for a pair of photons using a process called spontaneous parametric down-conversion.

You might suppose that this is because for half the pairs, the left atom is generated with $\mu_z = +\mu_B$ while the right atom is generated with $\mu_z = -\mu_B$, while for the other half of the pairs, the left atom is generated with $\mu_z = -\mu_B$ while the right atom is generated with $\mu_z = +\mu_B$. This supposition seems suspicious, because it singles out the z axis as special, but at this stage in our experimentation it's possible.

(2) Repeat the above experiment with horizontal Stern-Gerlach analyzers. The experimental result: Exactly the same as in experiment **(1)**! The two atoms always exit through opposite ports.

Problem 2.9 on page 83 demonstrates that the results of this experiment rule out the supposition presented at the bottom of experiment **(1)**.

(3) Repeat the above experiment with the two Stern-Gerlach analyzers oriented at $+120°$, or with both oriented at $-120°$, or with both oriented at $57°$, or for any other angle, as long as both have the same orientation. The experimental result: Exactly the same for any orientation!

(4) In an attempt to trick the atoms, we set the analyzers to vertical, then launch the pair of atoms, then (while the atoms are in flight) switch both analyzers to, say, $42°$, and have the atoms encounter these analyzers both with switched orientation. The experimental result: Regardless of what the orientation is, and regardless of when that orientation is set, the two atoms always exit through opposite ports.

Here is one way to picture this situation: The pair of atoms has a total magnetic moment of zero. But whenever the projection of a single atom on any axis is measured, the result must be $+\mu_B$ or $-\mu_B$, never zero. The only way to insure that that total magnetic moment, projected on any axis, sums to zero is the way described above. Do not put too much weight on this picture: like the "wants to go straight" story of section 2.1.4 (page 51), this is a classical story that happens to give the correct result. The definitive answer to any question is always experiment, not any picture or story, however appealing it may be.

These four experiments show that it is impossible to describe the condition of the atoms through anything like "the left atom has $\mu_z = +\mu_B$, the right atom has $\mu_z = -\mu_B$". How *can* we describe the condition of the pair? This will require further experimentation. For now, we say it has an EPR condition.

2.5.3 *EPR atom pair encounters flipping Stern-Gerlach analyzers*

A pair of atoms leaves the EPR source, and each atom travels at the same speed to vertical analyzers located 100 meters away. The left atom exits the − port, the right atom exits the + port. When the pair is flying from source to analyzer, it's not correct to describe it as "the left atom has $\mu_z = -\mu_B$, the right atom has $\mu_z = +\mu_B$", but after the atoms leave their analyzers, then this is a correct description.

Now shift the left analyzer one meter closer to the source. The left atom encounters its analyzer before the right atom encounters its. Suppose the left atom exits the − port, while the right atom is still in flight toward its analyzer. We know that when the right atom eventually does encounter its vertical analyzer, it will exit the + port. Thus it is correct to describe the right atom as having "$\mu_z = +\mu_B$", even though that atom hasn't yet encountered its analyzer.

Replace the right vertical analyzer with a flipping Stern-Gerlach analyzer. (In the figure below, it is in orientation O, out of the page.) Suppose the left atom encounters its vertical analyzer and exits the − port. Through the reasoning of the previous paragraph, the right atom now has $\mu_z = +\mu_B$. We know that when such an atom encounters a flipping Stern-Gerlach analyzer, it exits the + port with probability $\frac{1}{2}$.

Similarly, if the left atom encounters its vertical analyzer and exits the + port, the right atom now has $\mu_z = -\mu_B$, and once it arrives at its flipping analyzer, it will exit the − port with probability $\frac{1}{2}$. Summarizing these two paragraphs: Regardless of which port the left atom exits, the right atom will exit the opposite port with probability $\frac{1}{2}$.

Now suppose that the left analyzer were not vertical, but instead in orientation I, tilted into the page by one-third of a circle. It's easy to see that, again, regardless of which port the left atom exits, the right atom will exit the opposite port with probability $\frac{1}{2}$.

Finally, suppose that the left analyzer is a flipping analyzer. Once again, the two atoms will exit from opposite ports with probability $\frac{1}{2}$.

The above analysis supposed that the left analyzer was one meter closer to the source than the right analyzer, but clearly it also works if the right analyzer is one meter closer to the source than the left analyzer. Or one centimeter. One suspects that the same result will hold even if the two analyzers are exactly equidistant from the source, and experiment bears out this suspicion.

In summary: Each atom from this EPR source enters a flipping Stern-Gerlach analyzer.

(A) The atoms exit from opposite ports with probability $\frac{1}{2}$.
(B) If the two analyzers happen to have the same orientation, the atoms exit from opposite ports.

This is the prediction of quantum mechanics, and experiment confirms this prediction.

2.5.4 *The prediction of local determinism*

Suppose you didn't know anything about quantum mechanics, and you were told the result that "if the two analyzers have the same orientation, the atoms exit from opposite ports." Could you explain it?

I am sure you could. In fact, there are two possible explanations: First, the communication explanation. The left atom enters its vertical analyzer, and notices that it's being pulled toward the + port. It calls up the right atom with its walkie-talkie and says "If your analyzer has orientation I or O then you might go either way, but if your analyzer has orientation V you've got to go to the − port!" This is a possible explanation, but it's not a *local* explanation. The two analyzers might be 200 meters apart, or they might be 200 light-years apart. In either case, the message would have to get from the left analyzer to the right analyzer instantaneously. The walkie-talkies would have to use not radio waves, which propagate at the speed of light, but some sort of not-yet-discovered "insta-rays". Physicists have always been skeptical of non-local explanations, and since the advent of relativity they have grown even more skeptical, so we set this explanation aside. Can you find a local explanation?

Again, I am sure you can. Suppose that when the atoms are launched, they have some sort of characteristic that specifies which exit port they will take when they arrive at their analyzer. This very reasonable supposition, called "determinism", pervades all of classical mechanics. It is similar to saying "If I stand atop a 131 meter cliff and toss a ball horizontally with speed 23.3 m/s, I can predict the angle with which the ball strikes the ground, even though that event will happen far away and long in the future." In the case of the ball, the resulting strike angle is encoded into the initial position and velocity. In the case of the atoms, it's not clear how the exit port will be encoded: perhaps through the orientation of its magnetic moment, perhaps in some other, more elaborate way. But the method of encoding is irrelevant: if local determinism holds, then something within the atom determines which exit port it will take when it reaches its analyzer.[11] I'll represent this "something" through a code like $(++-)$. The first symbol means that if the atom encounters an analyzer in orientation V, it will exit through the $+$ port. The second means that if it encounters an analyzer in orientation O, it will exit through the $+$ port. The third means that if it encounters an analyzer in orientation I, it will exit through the $-$ port. The only way to ensure that "if the two analyzers have the same orientation, the atoms exit from opposite ports" is to assume that when the two atoms separate from each other within the source, they have opposite codes. If the left atom has $(+-+)$, the right atom must have $(-+-)$. If the left atom has $(---)$, the right atom must have $(+++)$. This is the local deterministic scheme for explaining fact (B) that "if the two analyzers have the same orientation, the atoms exit from opposite ports".

But can this scheme explain fact (A)? Let's investigate. Consider first the case mentioned above: the left atom has $(+-+)$ and the right atom has $(-+-)$. These atoms will encounter analyzers set to any of $3^2 = 9$ possible pairs of orientations. We list them below, along with with exit ports taken by the atoms. (For example, the third line of the table considers a left analyzer in orientation V and a right analyzer in orientation I. The left atom has code $(+-+)$, and the first entry in that code determines that the left atom will exit from the V analyzer through the $+$ port. The right atom has code $(-+-)$, and the third entry in that code determines that the right atom will exit from the I analyzer through the $-$ port.)

[11]But remember that in quantum mechanics determinism does not hold. The information *can't* be encoded within the three projections of a classical magnetic moment vector, because at any one instant, the quantum magnetic moment vector has only one projection.

left port	left analyzer	right analyzer	right port	opposite?
+	V	V	−	yes
+	V	O	+	no
+	V	I	−	yes
−	O	V	−	no
−	O	O	+	yes
−	O	I	−	no
+	I	V	−	yes
+	I	O	+	no
+	I	I	−	yes

Each of the nine orientation pairs (VV, OI, etc.) are equally likely, five of the orientation pairs result in atoms exiting from opposite ports, so when atoms of this type emerge from the source, the probability of these atoms exiting from opposite ports is $\frac{5}{9}$.

What about a pair of atoms generated with different codes? Suppose the left atom has $(--+)$ so the right atom must have $(++-)$. If you perform the analysis again, you will find that the probability of atoms exiting from opposite ports is once again $\frac{5}{9}$.

Suppose the left atom has $(---)$, so the right atom must have $(+++)$. The probability of the atoms exiting from opposite ports is of course 1.

There are, in fact, just $2^3 = 8$ possible codes:

code for left atom	probability of exiting opposite
$+++$	1
$-++$	5/9
$+-+$	5/9
$++-$	5/9
$+--$	5/9
$-+-$	5/9
$--+$	5/9
$---$	1

If the source makes left atoms of only type $(--+)$, then the probability of atoms exiting from opposite ports is $\frac{5}{9}$. If the source makes left atoms of only type $(+++)$, then the probability of atoms exiting from opposite ports is 1. If the source makes left atoms of type $(--+)$ half the time, and of type $(+++)$ half the time, then the probability of atoms exiting from opposite ports is halfway between $\frac{5}{9}$ and 1, namely $\frac{7}{9}$. But no matter how the source makes atoms, the probability of atoms exiting from opposite ports must be somewhere between $\frac{5}{9}$ and 1.

But experiment and quantum mechanics agree: That probability is actually $\frac{1}{2}$ — and $\frac{1}{2}$ is *not* between $\frac{5}{9}$ and 1. No local deterministic scheme — no matter how clever, or how elaborate, or how baroque — can give the result $\frac{1}{2}$. There is no "something within the atom that determines which exit port it will take when it reaches its analyzer". If the magnetic moment has a projection on axis V, then it *doesn't have* a projection on axis O or axis I.

There is a reason that Einstein, despite his many attempts, never produced a scheme that explained quantum mechanics in terms of some more fundamental, local and deterministic mechanism. It is not that Einstein wasn't clever. It is that *no such scheme exists.*

2.5.5 *The upshot*

This is a new phenomenon — one totally absent from classical physics — so it deserves a new name, something more descriptive than "EPR". Einstein called it "spooky action at a distance".[12] The phenomenon is spooky all right, but this phrase misses the central point that the phenomenon involves "correlations at a distance", whereas the word "action" suggests "cause-and-effect at a distance". Schrödinger coined the term "entanglement" for this phenomenon and said it was "not... *one* but rather *the* characteristic trait of quantum mechanics, the one that enforces its entire departure from classical lines of thought".[13] The world has followed Schrödinger and the phenomenon is today called entanglement. We will later investigate entanglement in more detail, but for now we will just call our EPR source a

[12]Letter from Einstein to Max Born, 3 March 1947, *The Born-Einstein Letters* (Macmillan, New York, 1971) translated by Irene Born.

[13]Erwin Schrödinger, "Discussion of probability relations between separated systems" *Mathematical Proceedings of the Cambridge Philosophical Society* **31** (October 1935) 555–563.

"source of entangled atom pairs" and describe the condition of the atom pair as "entangled".

The failure of local determinism described above is a special case of "Bell's Theorem", developed by John Bell[14] in 1964. The theorem has by now been tested experimentally numerous times in numerous contexts (various different angles; various distances between the analyzers; various sources of entangled pairs; various kinds of particles flying apart — gamma rays, or optical photons, or ions). In every test, quantum mechanics has been shown correct and local determinism wrong. What do we gain from these results?

First, they show that nature does not obey local determinism. To our minds, local determinism is common sense and any departure from it is weird. Thus whatever theory of quantum mechanics we eventually develop will be, to our eyes, weird. This will be a strength, not a defect, in the theory. The weirdness lies in nature, not in the theory used to describe nature.

Each of us feels a strong psychological tendency to reject the unfamiliar. In 1633, the Holy Office of the Inquisition found Galileo Galilei's idea that the Earth orbited the Sun so unfamiliar that they rejected it. The inquisitors put Galileo on trial and forced him to abjure his position. From the point of view of nature, the trial was irrelevant, Galileo's abjuration was irrelevant: the Earth orbits the Sun whether the Holy Office finds that fact comforting or not. It is our job as scientists to change our minds to fit nature; we do not change nature to fit our preconceptions. Don't make the inquisitors' mistake.

Second, the Bell's theorem result guides not just our calculations about nature but also our visualizations of nature, and even the very idea of what it means to "understand" nature. Lord Kelvin[15] framed the situation perfectly in his 1884 Baltimore lectures: "I never satisfy myself until I can

[14]John Stewart Bell (1928–1990), a Northern Irish physicist, worked principally in accelerator design, and his investigation of the foundations of quantum mechanics was something of a hobby. Concerning tests of his theorem, he remarked that "The reasonable thing just doesn't work." [Jeremy Bernstein, *Quantum Profiles* (Princeton University Press, Princeton, NJ, 1991) page 84.]

[15]William Thomson, the first Baron Kelvin (1824–1907), was an Irish mathematical physicist and engineer who worked in Scotland. He is best known today for establishing the thermodynamic temperature scale that bears his name, but he also made fundamental contributions to electromagnetism. He was knighted for his engineering work on the first transatlantic telegraph cable.

make a mechanical model of a thing. If I can make a mechanical model I can understand it. As long as I cannot make a mechanical model all the way through I cannot understand, and this is why I cannot get the electromagnetic theory."[16] If we take this as our meaning of "understand", then the experimental tests of Bell's theorem assure us that we will never be able to understand quantum mechanics.[17] What is to be done about this? There are only two choices. Either we can give up on understanding, or we can develop a new and more appropriate meaning for "understanding".

Max Born[18] argued for the first choice: "The ultimate origin of the difficulty lies in the fact (or philosophical principle) that we are compelled to use the words of common language when we wish to describe a phenomenon, not by logical or mathematical analysis, but by a picture appealing to the imagination. Common language has grown by everyday experience and can never surpass these limits."[19] Born felt that it was impossible to visualize or "understand" quantum mechanics: all you could do was grind through the "mathematical analysis".

Humans are visual animals, however, and I have found that when we are told not to visualize, we do so anyway. But we do so in an illicit and uncritical way. For example, many people visualize an atom passing through an interferometer as a small, hard, marble, with a definite position, despite the already-discovered fact that this visualization is untenable. Many people visualize a photon as a "ball of light" despite the fact that a photon (as conventionally defined) has a definite energy and hence can never have a position.

It *is* possible to develop a visualization and understanding of quantum mechanics. This *can't* be done by building a "mechanical model all the way through". It must be done through *both* analogy and contrast: atoms

[16]William Thomson, "Baltimore lectures on wave theory and molecular dynamics," in Robert Kargon and Peter Achinstein, editors, *Kelvin's Baltimore Lectures and Modern Theoretical Physics* (MIT Press, Cambridge, MA, 1987) page 206.

[17]The first time I studied quantum mechanics seriously, I wrote in the margin of my textbook "Good God they do it! But how?" I see now that I was looking for a mechanical mechanism undergirding quantum mechanics. It doesn't exist, but it's very natural for anyone to *want* it to exist.

[18]Max Born (1882–1970) was a German-Jewish theoretical physicist with a particular interest in optics. At the University of Göttingen in 1925 he directed Heisenberg's research which resulted in the first formulation of quantum mechanics. His granddaughter, the British-born Australian actress and singer Olivia Newton-John, is famous for her 1981 hit song "Physical".

[19]Max Born, *Atomic Physics*, sixth edition (Hafner Press, New York, 1957) page 97.

behave in *some* ways like small hard marbles, in *some* ways like classical waves, and in *some* ways like a cloud or fog of probability. Atoms don't behave *exactly* like any of these things, but if you keep in mind both the analogy and its limitations, then you can develop a pretty good visualization and understanding.

And that brings us back to the name "entanglement". It's an important name for an important phenomenon, but it suggests that the two distant atoms are connected mechanically, through strings. They aren't. The two atoms are correlated — if the left comes out $+$, the right comes out $-$, and vice versa — but they aren't correlated because of some signal sent back and forth through either strings or walkie-talkies. Entanglement involves correlation without causality.

Problems

2.8 **An atom walks into an analyzer**
Execute the "similar analysis" mentioned in the sentence below equation (2.2).

2.9 **A supposition squashed** (essential problem)
If atoms were generated according to the supposition presented below experiment **(1)** on page 74, then would would happen when they encountered the two horizontal analyzers of experiment **(2)**?

2.10 **A probability found through local determinism**
Suppose that the codes postulated on page 78 did exist. Suppose also that a given source produces the various possible codes with these probabilities:

code for left atom	probability of making such a pair
$+ + +$	1/2
$+ + -$	1/4
$+ - -$	1/8
$- - +$	1/8

If this given source were used in the experiment of section 2.5.3 with distant flipping Stern-Gerlach analyzers, what would be the probability of the two atoms exiting from opposite ports?

2.11 **A probability found through quantum mechanics**
In the test of Bell's inequality (the experiment of section 2.5.3), what is the probability given by quantum mechanics that, if the orientation settings are different, the two atoms exit from opposite ports?

2.6 Quantum cryptography

We've seen a lot of new phenomena, and the rest of this book is devoted to filling out our understanding of these phenomena and applying that understanding to various circumstances. But first, can we use them for anything?

We can. The sending of coded messages used to be the province of armies and spies and giant corporations, but today everyone does it. All transactions through automatic teller machines are coded. All Internet commerce is coded. This section describes a particular, highly reliable encoding scheme and then shows how quantal entanglement may someday be used to implement this scheme. (Quantum cryptography was used to securely transmit voting ballots cast in the Geneva canton of Switzerland during parliamentary elections held 21 October 2007. But it is not today in regular use anywhere.)

In this section I use names conventional in the field of coded messages (called cryptography). Alice and Bob wish to exchange private messages, but they know that Eve is eavesdropping on their communication. How can they encode their messages to maintain their privacy?

2.6.1 *The Vernam cipher*

The Vernam cipher or "one-time pad" technique is the only coding scheme proven to be absolutely unbreakable (if used correctly). It does not *rely* on the use of computers — it was invented by Gilbert Vernam in 1919 — but today it is mostly implemented using computers, so I'll describe it in that context.

Data are stored on computer disks through a series of magnetic patches on the disk that are magnetized either "up" or "down". An "up" patch is taken to represent `1`, and a "down" patch is taken to represent `0`. A string of seven patches is used to represent a character. For example, by a convention called ASCII, the letter "a" is represented through the sequence 1100001 (or, in terms of magnetizations, up, up, down, down, down, down, up). The letter "W" is represented through the sequence 1010111. Any computer the world around will represent the message "What?" through the sequence

1010111 1101000 1100001 1110100 0111111

This sequence is called the "plaintext".

But Alice doesn't want a message recognizable by any computer the world around. She wants to send the message "What?" to Bob in such a way that Eve will *not* be able to read the message, even though Eve has eavesdropped on the message. Here is the scheme invented by Vernam: Before sending her message, Alice generates a string of random 0s and 1s just as long as the message she wants to send — in this case, $7 \times 5 = 35$ bits. She might do this by flipping 35 coins, or by flipping one coin 35 times. I've just done that, producing the random number

0100110 0110011 1010110 1001100 1011100

Then Alice gives Bob a copy of that random number – the "key".

Instead of sending the plaintext, Alice modifies her plaintext into a coded "ciphertext" using the key. She writes down her plaintext and writes the key below it, then works through column by column. For each position, if the key is 0 the plaintext is left unchanged; but if the key is 1 the plaintext is reversed (from 0 to 1 or vice versa). For the first column, the key is 0, so Alice *doesn't* change the plaintext: the first character of ciphertext is the same as the first character of plaintext. For the second column, the key is 1, so Alice *does* change the plaintext: the second character of ciphertext is the reverse of the second character of plaintext. Alice goes through all the columns, duplicating the plaintext where the key is 0 and reversing the plaintext where the key is 1.

plaintext:	1010111 1101000 1100001 1110100 0111111
key:	0100110 0110011 1010110 1001100 1011100
ciphertext:	1110001 1011011 0110111 0111000 1100011

Then, Alice sends out her ciphertext over open communication lines.

Now, the ciphertext that Bob (and Eve) receive translates to *some* message through the ASCII convention – in fact, it translates to "q[78c" — but because the key is random, the ciphertext is just as random. Bob deciphers Alice's message by carrying out the encoding process on the ciphertext, namely, duplicating the ciphertext where the key is 0 and reversing the ciphertext where the key is 1. The result is the plaintext. Eve does not know the key, so she cannot produce the plaintext.

The whole scheme relies on the facts that the key is (1) random and (2) unknown to Eve. The very name "one-time pad" underscores that a key can only be used once and must then be discarded. If a single key is used for two messages, then the second key is not "random" — it is instead perfectly correlated with the first key. There are easy methods to break the code when a key is reused.

Generating random numbers is not easy, and the Vernam cipher demands keys as long as the messages transmitted. As recently as 1992, high-quality computer random-number generators were classified by the U.S. government as munitions, along with tanks and fighter planes, and their export from the country was prohibited.

And of course Eve must not know the key. So there must be some way for Alice to get the key to Bob securely. If they have some secure method for transmitting keys, why don't they just use that same secure method for sending their messages?

In common parlance, the word "random" can mean "unimportant, not worth considering" (as in "Joe made a random comment"). So it may seem remarkable that a major problem for government, the military, and commerce is the generation and distribution of randomness, but that is indeed the case.

2.6.2 *Quantum mechanics to the rescue*

Since quantum mechanics involves randomness, it seems uniquely positioned to solve this problem. Here's one scheme.

Alice and Bob set up a source of entangled atoms halfway between their two homes. Both of them erect vertical Stern-Gerlach analyzers to detect the atoms. If Alice's atom comes out +, she will interpret it as a 1, if −, a 0. Bob interprets his atoms in the opposite sense. Since the entangled atoms always exit from opposite ports, Alice and Bob end up with the same random number, which they use as a key for their Vernam-cipher communications over conventional telephone or computer lines.

This scheme will indeed produce and distribute copious, high-quality random numbers. But Eve can get at those same numbers through the following trick: She cuts open the atom pipe leading from the entangled source to Alice's home, and inserts a vertical interferometer.[20] She watches the atoms pass through her interferometer. If the atom takes path a, Eve knows that when Alice receives that same atom, it will exit from Eve's + port. If the atom takes path b, the opposite holds. Eve gets the key, Eve breaks the code.

It's worth looking at this eavesdropping in just a bit more detail. When the two atoms depart from their source, they are entangled. It is *not* true that, say, Alice's atom has $\mu_z = +\mu_B$ while Bob's atom has $\mu_z = -\mu_B$ — the pair of atoms is in the condition we've called "entangled", but the individual atoms themselves are not in *any* condition. However, after Eve sees the atom taking path a of her interferometer, then the two atoms are no longer entangled — now it *is* true that Alice's atom has the condition $\mu_z = +\mu_B$ while Bob's atom has the condition $\mu_z = -\mu_B$. The key received by Alice and Bob will be random whether or not Eve is listening in. To test for evesdropping, Alice and Bob must examine it in some other way.

Replace Alice and Bob's vertical analyzers with flipping Stern-Gerlach analyzers. After Bob receives his random sequence of pluses and minuses, he sends it to Alice over an open communication line. (Eve will intercept that sequence but it won't do her any good, because Bob sends only the pluses and minuses, not the orientations of his analyzer.) Alice now knows both the results at her analyzer and the results at Bob's analyzer, so she can perform a test of Bell's theorem: If she finds that the probability of atoms coming out opposite is $\frac{1}{2}$, then she knows that their atoms have arrived entangled, thus Eve has *not* observed the atoms in transit. If she finds that the probability is between $\frac{5}{9}$ and 1, then she knows for certain that Eve *is* listening in, and they must not use their compromised key.

Is there some other way for Eve to tap the line? No! If the atom pairs pass the test for entanglement, then no one can know the values of their

[20]Inspired by James Bond, I always picture Eve as exotic beauty in a little black dress slinking to the back of an eastern European café to tap the diplomatic cable which conveniently runs there. But in point of fact Eve would be a computer.

μ_z projections because those projections don't exist! We have guaranteed that no one has intercepted the key by the interferometer method, or by any other method whatsoever.

Once Alice has tested Bell's theorem, she and Bob still have a lot of work to do. For a key they must use only those random numbers produced when their two analyzers happen to have the same orientations. There are detailed protocols specifying how Alice and Bob must exchange information about their analyzer orientations, in such a way that Eve can't uncover them. I won't describe these protocols because while they tell you how clever people are, they tell you nothing about how nature behaves. But you should take away that entanglement is not merely a phenomenon of nature: it is also a natural resource.

2.7 What is a qubit?

We've devoted an entire chapter to the magnetic moment of a silver atom. Perhaps you find this inappropriate: do you really care so much about silver atoms? Yes you do, because the phenomena and principles we've established concerning the magnetic moment of a silver atom apply to a host of other systems: the polarization of a light photon, the hybridization of a benzene molecule, the position of the nitrogen atom within an ammonia molecule, the neutral kaon, and more. Such systems are called "two-state systems" or "spin-$\frac{1}{2}$ systems" or "qubit systems". The ideas we establish concerning the magnetic moment of a silver atom apply equally well to all these systems.

After developing these ideas in the next chapter, we will (in chapter 4, "The Quantum Mechanics of Position") generalize them to continuum systems like the position of an electron.

Problem

2.12 **Questions** (recommended problem)
Update your list of quantum mechanics questions that you started at problem 1.17 on page 46. Write down new questions and, if you have uncovered answers to any of your old questions, write them down briefly.

Chapter 3

Forging Mathematical Tools

When you walked into your introductory classical mechanics course, you were already familiar with the phenomena of introductory classical mechanics: flying balls, spinning wheels, colliding billiard balls. Your introductory mechanics textbook didn't need to introduce these things to you, but instead jumped right into describing these phenomena mathematically and explaining them in terms of more general principles.

The last chapter of this book made you familiar with the phenomena of quantum mechanics: quantization, interference, and entanglement — at least, insofar as these phenomena are manifest in the behavior of the magnetic moment of a silver atom. You are now, with respect to quantum mechanics, at the same level that you were, with respect to classical mechanics, when you walked into your introductory mechanics course. It is now our job to describe these quantal phenomena mathematically, to explain them in terms of more general principles, and (eventually) to investigate situations more complex than the magnetic moment of one or two silver atoms.

3.1 What is a quantal state?

We've been talking about the state of the silver atom's magnetic moment by saying things like "the projection of the magnetic moment on the z axis is $\mu_z = -\mu_B$" or "$\mu_x = +\mu_B$" or "$\mu_\theta = -\mu_B$". This notation is clumsy. First of all, it requires you to write down the same old μs time and time again. Second, the most important thing is the axis (z or x or θ), and the symbol for the axis is also the smallest and easiest to overlook.

P.A.M. Dirac[1] invented a notation that overcomes these faults. He looked at descriptions like

$$\mu_z = -\mu_B \quad \text{or} \quad \mu_x = +\mu_B \quad \text{or} \quad \mu_\theta = -\mu_B$$

and noted that the only difference from one expression to the other was the axis subscript and the sign in front of μ_B. Since the only thing that distinguishes one expression from another is $(z, -)$, or $(x, +)$, or $(\theta, -)$, Dirac thought, these should be the only things we need to write down. He denoted these three states as

$$|z-\rangle \quad \text{or} \quad |x+\rangle \quad \text{or} \quad |\theta-\rangle.$$

The placeholders $|\quad\rangle$ are simply ornaments to remind us that we're talking about quantal states, just as the arrow atop $\vec{r}$ is simply an ornament to remind us that we're talking about a vector. States expressed using this notation are sometimes called "kets".

Simply establishing a notation doesn't tell us much. Just as in classical mechanics, we say we know a state when we know all the information needed to describe the system now and to predict its future. In our universe the classical time evolution law is

$$\vec{F} = m\frac{d^2\vec{r}}{dt^2}$$

and so the state is specified by giving both a position $\vec{r}$ and a velocity $\vec{v}$. If nature had instead provided the time evolution law

$$\vec{F} = m\frac{d^3\vec{r}}{dt^3}$$

then the state would have been specified by giving a position $\vec{r}$, a velocity $\vec{v}$, and an acceleration $\vec{a}$. The specification of state is dictated by nature, not by humanity, so we can't know how to specify a state until we know the laws of physics governing that state. Since we don't yet know the laws of quantal physics, we can't yet know exactly how to specify a quantal state.

Classical intuition makes us suppose that, to specify the magnetic moment of a silver atom, we need to specify all three components μ_z, μ_x, and μ_y. We have already seen that nature precludes such a specification: if the magnetic moment has a value for μ_z, then it doesn't have a value for μ_x,

[1]The Englishman Paul Adrien Maurice Dirac (1902–1984) in 1928 formulated a relativistically correct quantum mechanical equation that turns out to describe the electron. In connection with this so-called Dirac equation, he predicted the existence of antimatter. Dirac was painfully shy and notoriously cryptic.

and it's absurd to demand a specification for something that doesn't exist. As we learn more and more quantum physics, we will learn better and better how to specify states. There will be surprises. But always keep in mind that (just as in classical mechanics) it is experiment, not philosophy or meditation, and certainly not common sense, that tells us how to specify states.

3.2 Amplitude

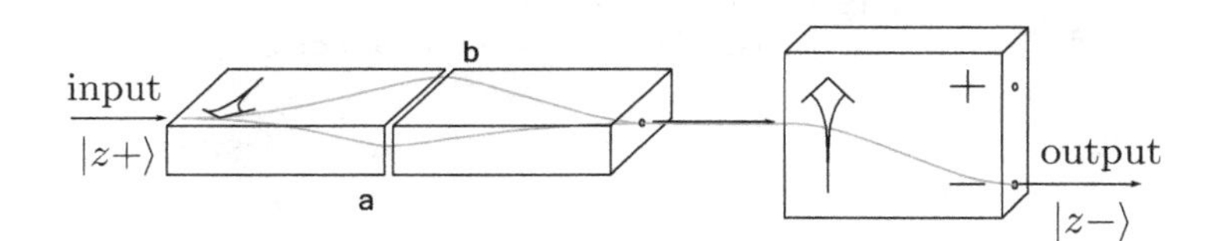

An atom in state $|z+\rangle$ ambivates through the apparatus above. We have already seen that, when the atom ambivates in darkness,

$$\begin{aligned}&\text{probability to go from input to output} \neq \\ &\quad \text{probability to go from input to output via path a} \\ &\quad + \text{probability to go from input to output via path b.}\end{aligned} \tag{3.1}$$

On the other hand, it makes sense to associate some sort of "influence to go from input to output via path a" with the path through a and a corresponding "influence to go from input to output via path b" with the path through b. This postulated influence is called "probability amplitude" or just "amplitude".[2] Whatever amplitude is, its desired property is that

$$\begin{aligned}&\text{amplitude to go from input to output} = \\ &\quad \text{amplitude to go from input to output via path a} \\ &\quad + \text{amplitude to go from input to output via path b.}\end{aligned} \tag{3.2}$$

For the moment, the very existence of amplitude is nothing but a hopeful surmise. Scientists cannot now and indeed never will be able to prove that the concept of amplitude applies to all situations. That's because new situations are being investigated every day, and perhaps tomorrow a new

[2]The name "amplitude" is a poor one, because it is also used for the maximum value of a sinusoidal signal — in the function $A\sin(\omega t)$, the symbol A represents the amplitude — and this sinusoidal signal "amplitude" has nothing to do with the quantal "amplitude". One of my students correctly suggested that a better name for quantal amplitude would be "proclivity". But it's too late now to change the word.

situation will be discovered that cannot be described in terms of amplitudes. But as of today, that hasn't happened.

The role of amplitude, whatever it may prove to be, is to calculate probabilities. We establish three desirable rules:

(1) *From amplitude to probability.* For every possible action there is an associated amplitude, such that

$$\text{probability for the action} = |\text{amplitude for the action}|^2.$$

(2) *Actions in series.* If an action takes place through several successive stages, the amplitude for that action is the product of the amplitudes for each stage.
(3) *Actions in parallel.* If an action could take place in several possible ways, the amplitude for that action is the sum of the amplitudes for each possibility.

The first rule is a simple way to make sure that probabilities are always positive. The second rule is a natural generalization of the rule for probabilities in series — that if an action happens through several stages, the probability for the action as a whole is the product of the probabilities for each stage. And the third rule simply restates the "desired property" presented in equation (3.2).

We apply these rules to various situations that we've already encountered, beginning with the interference experiment sketched above. Recall the probabilities already established (first column in table):

	probability	|amplitude|	amplitude
go from input to output	0	0	0
go from input to output via path a	$\frac{1}{4}$	$\frac{1}{2}$	$+\frac{1}{2}$
go from input to output via path b	$\frac{1}{4}$	$\frac{1}{2}$	$-\frac{1}{2}$

If rule (1) is to hold, then the amplitude to go from input to output must also be 0, while the amplitude to go via a path must have magnitude $\frac{1}{2}$ (second column in table). According to rule (3), the two amplitudes to go via a and via b must sum to zero, so they cannot both be represented by positive numbers. Whatever mathematical entity is used to represent amplitude, it must enable two such entities, each with non-zero magnitude, to sum to zero. There are many such entities: real numbers, complex numbers, hypercomplex numbers, and vectors in three dimensions are all possibilities. For this particular interference experiment, it suffices to assign real numbers to amplitudes: the amplitude to go via path a is $+\frac{1}{2}$, and the amplitude to go via path b is $-\frac{1}{2}$. (Third column in table. The negative sign could have been assigned to path a rather than to path b: this choice is merely conventional.) For other interference experiments complex numbers are required. It turns out that, for all situations yet encountered, one can represent amplitude mathematically as a complex number. Once again, this reflects the results of experiment, not of philosophy or meditation.

The second situation we'll consider is a Stern-Gerlach analyzer.

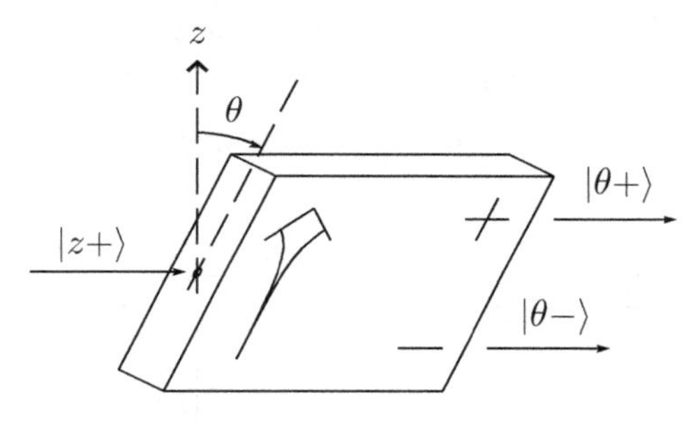

The amplitude that an atom entering the θ-analyzer in state $|z+\rangle$ exits in state $|\theta+\rangle$ is called[3] $\langle\theta+|z+\rangle$. That phrase is a real mouthful, so the symbol $\langle\theta+|z+\rangle$ is pronounced "the amplitude that $|z+\rangle$ is in $|\theta+\rangle$", even though this briefer pronunciation leaves out the important role of the analyzer.[4] From rule (1), we know that

$$|\langle\theta+|z+\rangle|^2 = \cos^2(\theta/2) \tag{3.3}$$

$$|\langle\theta-|z+\rangle|^2 = \sin^2(\theta/2). \tag{3.4}$$

You can also use rule (1), in connection with the experiments described in

[3]The states appear in the symbol in the opposite sequence from their appearance in the description.

[4]The ultimate source of such problems is that the English language was invented by people who did not understand quantum mechanics, hence they never produced concise, accurate phrases to describe quantal phenomena. In the same way, the ancient phrase "search the four corners of the Earth" is still colorful and practical, and is used today even by those who know that the Earth doesn't have four corners.

problem 2.1, "Exit probabilities" (on page 57) to determine that

$$\begin{aligned}
|\langle z+|\theta+\rangle|^2 &= \cos^2(\theta/2)\\
|\langle z-|\theta+\rangle|^2 &= \sin^2(\theta/2)\\
|\langle \theta+|z-\rangle|^2 &= \sin^2(\theta/2)\\
|\langle \theta-|z-\rangle|^2 &= \cos^2(\theta/2)\\
|\langle z+|\theta-\rangle|^2 &= \sin^2(\theta/2)\\
|\langle z-|\theta-\rangle|^2 &= \cos^2(\theta/2).
\end{aligned}$$

Clearly analyzer experiments like these find the *magnitude* of an amplitude. No analyzer experiment can find the *phase* of an amplitude.[5] To determine phases, we must perform interference experiments.

So the third situation is an interference experiment.

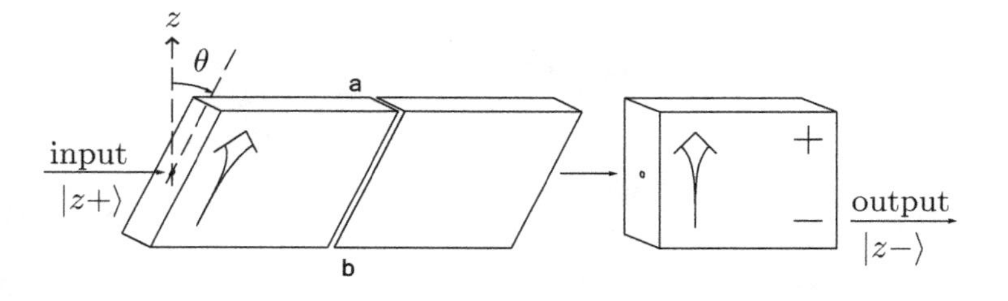

Rule (2), actions in series, tells us that the amplitude to go from $|z+\rangle$ to $|z-\rangle$ via path a is the product of the amplitude to go from $|z+\rangle$ to $|\theta+\rangle$ times the amplitude to go from $|\theta+\rangle$ to $|z-\rangle$:

$$\text{amplitude to go via path a} = \langle z-|\theta+\rangle\langle\theta+|z+\rangle.$$

Similarly

$$\text{amplitude to go via path b} = \langle z-|\theta-\rangle\langle\theta-|z+\rangle.$$

And then rule (3), actions in parallel, tells us that the amplitude to go from $|z+\rangle$ to $|z-\rangle$ is the sum of the amplitude to go via path a and the amplitude to go via path b. In other words

$$\langle z-|z+\rangle = \langle z-|\theta+\rangle\langle\theta+|z+\rangle + \langle z-|\theta-\rangle\langle\theta-|z+\rangle. \tag{3.5}$$

[5]The terms phase and magnitude are explained in appendix C, "Complex Arithmetic".

We know the magnitude of each of these amplitudes from analyzer experiments:

amplitude	magnitude
$\langle z- \mid z+\rangle$	0
$\langle z- \mid \theta+\rangle$	$\lvert\sin(\theta/2)\rvert$
$\langle \theta+ \mid z+\rangle$	$\lvert\cos(\theta/2)\rvert$
$\langle z- \mid \theta-\rangle$	$\lvert\cos(\theta/2)\rvert$
$\langle \theta- \mid z+\rangle$	$\lvert\sin(\theta/2)\rvert$

The task now is to assign phases to these magnitudes in such a way that equation (3.5) is satisfied. In doing so we are faced with an embarrassment of riches: there are *many* consistent ways to make this assignment. Here are two commonly used conventions:

amplitude	convention I	convention II
$\langle z- \mid z+\rangle$	0	0
$\langle z- \mid \theta+\rangle$	$\sin(\theta/2)$	$i\sin(\theta/2)$
$\langle \theta+ \mid z+\rangle$	$\cos(\theta/2)$	$\cos(\theta/2)$
$\langle z- \mid \theta-\rangle$	$\cos(\theta/2)$	$\cos(\theta/2)$
$\langle \theta- \mid z+\rangle$	$-\sin(\theta/2)$	$-i\sin(\theta/2)$

There are two things to notice about these amplitude assignments. First, one normally assigns values to physical quantities by experiment, or by calculation, but not "by convention". Second, both of these conventions show unexpected behaviors: Because the angle 0° is the same as the angle 360°, one would expect that $\langle 0^\circ+ \mid z+\rangle$ would equal $\langle 360^\circ+ \mid z+\rangle$, whereas in fact the first amplitude is $+1$ and the second is -1. Because the state $\lvert 180^\circ-\rangle$ (that is, $\lvert\theta-\rangle$ with $\theta = 180^\circ$) is the same as the state $\lvert z+\rangle$, one would expect that $\langle 180^\circ- \mid z+\rangle = 1$, whereas in fact $\langle 180^\circ- \mid z+\rangle$ is either -1 or $-i$, depending on convention. These two observations underscore the fact that amplitude is a mathematical tool that enables us to calculate physically observable quantities, like probabilities. It is not itself a physical entity. No experiment measures amplitude. Amplitude is not "out there, physically present in space" in the way that, say, a nitrogen molecule is.

A good analogy is that an amplitude convention is like a language. Any language is a human convention: there is no intrinsic connection between a physical horse and the English word "horse", or the German word "pferd", or the Swahili word "farasi". The fact that language is pure human convention, and that there are multiple conventions for the name of a horse, doesn't mean that language is unimportant: on the contrary language is an immensely powerful tool. And the fact that language is pure human convention doesn't mean that you can't develop intuition about language: on the contrary if you know the meaning of "arachnid" and the meaning of "phobia", then your intuition for English tells you that "arachnophobia" means fear of spiders. Exactly the same is true for amplitude: it is a powerful tool, and with practice you can develop intuition for it.

When I introduced the phenomenon of quantal interference on page 62, I said that there was no word or phrase in the English language that accurately represents what's going on: It's flat-out wrong to say "the atom takes path a" and it's flat-out wrong to say "the atom takes path b". It gives a wrong impression to say "the atom takes no path" or "the atom takes both paths". I introduced the phrase "the atom ambivates through the two paths of the interferometer". Now we have a technically correct way of describing the phenomenon: "the atom has an amplitude to take path a and an amplitude to take path b".

Here's another warning about language: If an atom in state $|\psi\rangle$ enters a vertical analyzer, the amplitude for it to exit from the + port is $\langle z+|\psi\rangle$. (And of course the amplitude for it exit from the − port is $\langle z-|\psi\rangle$.) This is often stated "If the atom is in state $|\psi\rangle$, the amplitude of it being in state $|z+\rangle$ is $\langle z+|\psi\rangle$." This is an acceptable shorthand for the full explanation, which requires thinking about an analyzer experiment, even though the shorthand never mentions the analyzer. But never say "If the atom is in state $|\psi\rangle$, the probability of it being in state $|z+\rangle$ is $|\langle z+|\psi\rangle|^2$." This gives the distinct and incorrect impression that before entering the analyzer, the atom was either in state $|z+\rangle$ or in state $|z-\rangle$, and you just didn't know which it was. Instead, say "If an atom in state $|\psi\rangle$ enters a vertical analyzer, the probability of exiting from the + port in state $|z+\rangle$ is $|\langle z+|\psi\rangle|^2$."

3.2.1 *Sample Problem: Two paths*

Find an equation similar to equation (3.5) representing the amplitude to start in state $|\psi\rangle$ at input, ambivate through a vertical interferometer, and end in state $|\phi\rangle$ at output.

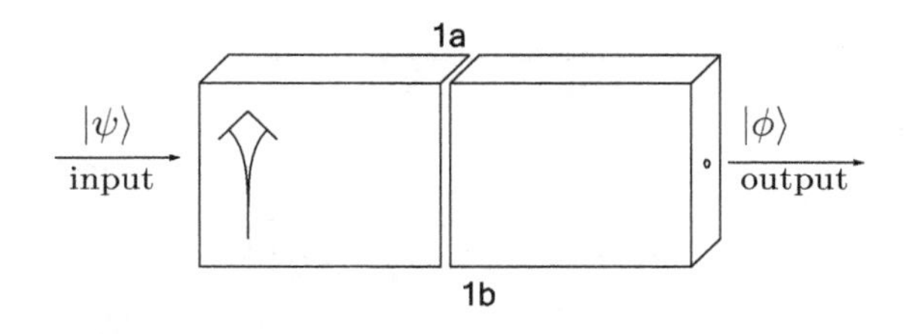

Solution: Because of rule (2), actions in series, the amplitude for the atom to take the top path is the product

$$\langle\phi|z+\rangle\langle z+|\psi\rangle.$$

Similarly the amplitude for it to take the bottom path is

$$\langle\phi|z-\rangle\langle z-|\psi\rangle.$$

Because of rule (3), actions in parallel, the amplitude for it to ambivate through both paths is the sum of these two, and we conclude that

$$\langle\phi|\psi\rangle = \langle\phi|z+\rangle\langle z+|\psi\rangle + \langle\phi|z-\rangle\langle z-|\psi\rangle. \tag{3.6}$$

3.2.2 *Sample Problem: Three paths*

Stretch apart a vertical interferometer, so that the recombining rear end is far from the splitting front end, and insert a θ interferometer into the bottom path. Now there are *three* paths from input to output. Find an equation similar to equation (3.5) representing the amplitude to start in state $|\psi\rangle$ at input and end in state $|\phi\rangle$ at output.

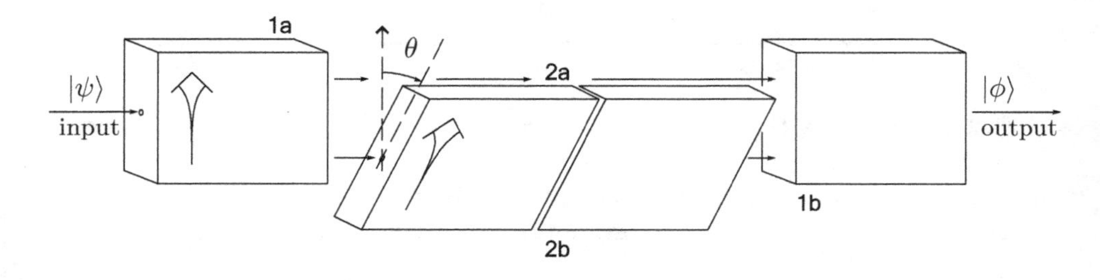

Solution:

$$\begin{aligned}\langle\phi|\psi\rangle = &\ \langle\phi|z+\rangle\langle z+|\psi\rangle \\ &+ \langle\phi|z-\rangle\langle z-|\theta+\rangle\langle\theta+|z-\rangle\langle z-|\psi\rangle \\ &+ \langle\phi|z-\rangle\langle z-|\theta-\rangle\langle\theta-|z-\rangle\langle z-|\psi\rangle\end{aligned} \tag{3.7}$$

Problems

3.1 **Talking about interference**
An atom in state $|\psi\rangle$ ambivates through a vertical analyzer. We say, appropriately, that "the atom has an amplitude to take the top path and an amplitude to take the bottom path". For the benefit of students in next year's offering of this class (see page 20), find expressions for those two amplitudes and describe, in ten sentences or fewer, why it is *not* appropriate to say "the atom has probability $|\langle z+|\psi\rangle|^2$ to take the top path and probability $|\langle z-|\psi\rangle|^2$ to take the bottom path".

3.2 **Other conventions**
Two conventions for assigning amplitudes are given in the table on page 95. Show that if $\langle z-|\theta+\rangle$ and $\langle z-|\theta-\rangle$ are multiplied by phase factor $e^{i\alpha}$, and if $\langle z+|\theta+\rangle$ and $\langle z+|\theta-\rangle$ are multiplied by phase factor $e^{i\beta}$ (where α and β are both real), then the resulting amplitudes are just as good as the original (for either convention I or convention II).

3.3 **Peculiarities of amplitude**
Page 95 pointed out some of the peculiarities of amplitude; this problem points out another. Since the angle θ is the same as the angle $360° + \theta$, one would expect that $\langle\theta+|z+\rangle$ would equal $\langle(360° + \theta)+|z+\rangle$. Show, using either of the conventions given in the table on page 95, that this expectation is false. What is instead correct?

3.3 Reversal-conjugation relation

The "complex conjugate" of any complex number is the same number but with every "i" changed to "$-i$": if x and y are real numbers, then

$$z = x + iy \quad \text{has complex conjugate} \quad z^* = x - iy. \tag{3.8}$$

A useful theorem says that the amplitude to go from state $|\psi\rangle$ to state $|\phi\rangle$ and the amplitude to go in the opposite direction are related through complex conjugation:

$$\langle\phi|\psi\rangle = \langle\psi|\phi\rangle^*. \tag{3.9}$$

The proof below works for states of the magnetic moment of a silver atom — the kind of states we've worked with so far — but in fact the result holds for any quantal system.

The proof relies on three facts: First, the probability for one state to be analyzed into another depends only on the magnitude of the angle between the incoming magnetic moment and the analyzer, and not on the sense of that angle. (An atom in state $|z+\rangle$ has the same probability of leaving the + port of an analyzer whether it is rotated 17° clockwise or 17° counterclockwise.) Thus

$$|\langle\phi|\psi\rangle|^2 = |\langle\psi|\phi\rangle|^2. \tag{3.10}$$

Second, an atom exits an interferometer in the same state in which it entered, so

$$\langle\phi|\psi\rangle = \langle\phi|\theta+\rangle\langle\theta+|\psi\rangle + \langle\phi|\theta-\rangle\langle\theta-|\psi\rangle. \tag{3.11}$$

Third, an atom entering an analyzer comes out somewhere, so

$$1 = |\langle\theta+|\psi\rangle|^2 + |\langle\theta-|\psi\rangle|^2. \tag{3.12}$$

The proof also relies on a mathematical result called "the triangle inequality for complex numbers": If a and b are real numbers with $a+b=1$, and in addition $e^{i\alpha}a + e^{i\beta}b = 1$, with α and β real, then $\alpha = \beta = 0$. You can find very general, very abstract, proofs of the triangle inequality, but the complex plane sketch below encapsulates the idea:

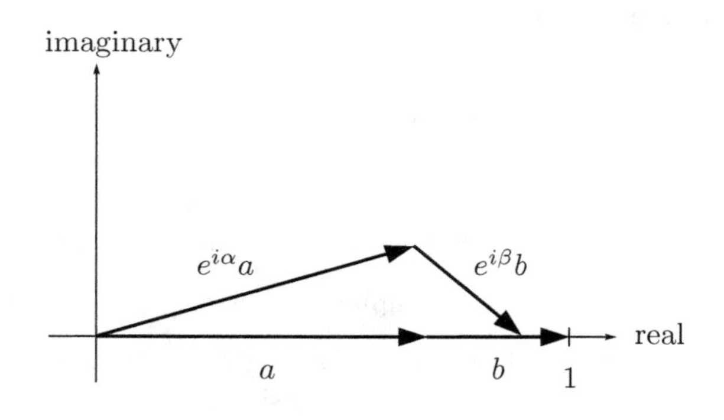

From the first fact (3.10), the two complex numbers $\langle\phi|\psi\rangle$ and $\langle\psi|\phi\rangle$ have the same magnitude, so they differ only in phase. Write this statement as

$$\langle\phi|\psi\rangle = e^{i\delta}\langle\psi|\phi\rangle^* \tag{3.13}$$

where the phase δ is a real number that might depend on the states $|\phi\rangle$ and $|\psi\rangle$. Apply this general result first to the particular state $|\phi\rangle = |\theta+\rangle$:

$$\langle\theta+|\psi\rangle = e^{i\delta_+}\langle\psi|\theta+\rangle^*, \tag{3.14}$$

and then to the particular state $|\phi\rangle = |\theta-\rangle$:

$$\langle\theta-|\psi\rangle = e^{i\delta_-}\langle\psi|\theta-\rangle^*, \tag{3.15}$$

where the two real numbers δ_+ and δ_- might be different. Our objective is to prove that $\delta_+ = \delta_- = 0$.

Apply the second fact (3.11) with $|\phi\rangle = |\psi\rangle$, giving

$$\begin{aligned} 1 &= \langle\psi|\theta+\rangle\langle\theta+|\psi\rangle + \langle\psi|\theta-\rangle\langle\theta-|\psi\rangle \\ &= e^{i\delta_+}\langle\psi|\theta+\rangle\langle\psi|\theta+\rangle^* + e^{i\delta_-}\langle\psi|\theta-\rangle\langle\psi|\theta-\rangle^* \\ &= e^{i\delta_+}|\langle\psi|\theta+\rangle|^2 + e^{i\delta_-}|\langle\psi|\theta-\rangle|^2 \\ &= e^{i\delta_+}|\langle\theta+|\psi\rangle|^2 + e^{i\delta_-}|\langle\theta-|\psi\rangle|^2. \end{aligned} \tag{3.16}$$

Compare this result to the third fact (3.12)

$$1 = |\langle\theta+|\psi\rangle|^2 + |\langle\theta-|\psi\rangle|^2 \tag{3.17}$$

and use the triangle inequality with $a = |\langle\theta+|\psi\rangle|^2$ and $b = |\langle\theta-|\psi\rangle|^2$. The two phases δ_+ and δ_- must vanish, so the "reversal-conjugation relation" is proven.

3.4 Establishing a phase convention

Although there are multiple alternative phase conventions for amplitudes (see problem 3.2 on page 98), we will from now on use only phase convention I from page 95:

$$\begin{aligned} \langle z+|\theta+\rangle &= \cos(\theta/2) \\ \langle z-|\theta+\rangle &= \sin(\theta/2) \\ \langle z+|\theta-\rangle &= -\sin(\theta/2) \\ \langle z-|\theta-\rangle &= \cos(\theta/2) \end{aligned} \tag{3.18}$$

In particular, for $\theta = 90^\circ$ we have

$$\begin{aligned} \langle z+|x+\rangle &= 1/\sqrt{2} \\ \langle z-|x+\rangle &= 1/\sqrt{2} \\ \langle z+|x-\rangle &= -1/\sqrt{2} \\ \langle z-|x-\rangle &= 1/\sqrt{2} \end{aligned} \tag{3.19}$$

This convention has a desirable special case for $\theta = 0^\circ$, namely

$$\begin{aligned} \langle z+|\theta+\rangle &= 1 \\ \langle z-|\theta+\rangle &= 0 \\ \langle z+|\theta-\rangle &= 0 \\ \langle z-|\theta-\rangle &= 1 \end{aligned} \tag{3.20}$$

but an unexpected special case for $\theta = 360^\circ$, namely

$$\begin{aligned} \langle z+|\theta+\rangle &= -1 \\ \langle z-|\theta+\rangle &= 0 \\ \langle z+|\theta-\rangle &= 0 \\ \langle z-|\theta-\rangle &= -1 \end{aligned} \tag{3.21}$$

This is perplexing, given that the angle $\theta = 0^\circ$ is the same as the angle $\theta = 360^\circ$! Any convention will have similar perplexing cases. Such perplexities underscore the fact that amplitudes are important mathematical tools used to calculate probabilities, but are not "physically real".

Given these amplitudes, we can use the interference result (3.6) to calculate any amplitude of interest:

$$\begin{aligned} \langle \phi|\psi\rangle &= \langle \phi|z+\rangle\langle z+|\psi\rangle + \langle \phi|z-\rangle\langle z-|\psi\rangle \\ &= \langle z+|\phi\rangle^*\langle z+|\psi\rangle + \langle z-|\phi\rangle^*\langle z-|\psi\rangle \end{aligned} \tag{3.22}$$

where in the last line we have used the reversal-conjugation relation (3.9).

Problems

3.4 Other conventions, other peculiarities

Write what this section would have been had we adopted convention II rather than convention I from page 95. In addition, evaluate the four amplitudes of equation (3.18) for $\theta = +180°$ and $\theta = -180°$.

3.5 Finding amplitudes (recommended problem)

Using the interference idea embodied in equation (3.22), calculate the amplitudes $\langle\theta{+}|54°{+}\rangle$ and $\langle\theta{-}|54°{+}\rangle$ as a function of θ. Do these amplitudes have the values you expect for $\theta = 54°$? For $\theta = 234°$? Plot $\langle\theta{+}|54°{+}\rangle$ for θ from 0° to 360°. Compare the result for $\theta = 0°$ and $\theta = 360°$.

3.6 Rotations

Use the interference idea embodied in equation (3.22) to show that

$$\begin{aligned}
\langle x{+}|\theta{+}\rangle &= \tfrac{1}{\sqrt{2}}[\cos(\theta/2) + \sin(\theta/2)] \\
\langle x{-}|\theta{+}\rangle &= -\tfrac{1}{\sqrt{2}}[\cos(\theta/2) - \sin(\theta/2)] \\
\langle x{+}|\theta{-}\rangle &= \tfrac{1}{\sqrt{2}}[\cos(\theta/2) - \sin(\theta/2)] \\
\langle x{-}|\theta{-}\rangle &= \tfrac{1}{\sqrt{2}}[\cos(\theta/2) + \sin(\theta/2)]
\end{aligned} \tag{3.23}$$

If and only if you enjoy trigonometric identities, you should then show that these results can be written equivalently as

$$\begin{aligned}
\langle x{+}|\theta{+}\rangle &= \cos((\theta - 90°)/2) \\
\langle x{-}|\theta{+}\rangle &= \sin((\theta - 90°)/2) \\
\langle x{+}|\theta{-}\rangle &= -\sin((\theta - 90°)/2) \\
\langle x{-}|\theta{-}\rangle &= \cos((\theta - 90°)/2)
\end{aligned} \tag{3.24}$$

This makes perfect geometric sense, as the angle relative to the x axis is 90° less than the angle relative to the z axis:

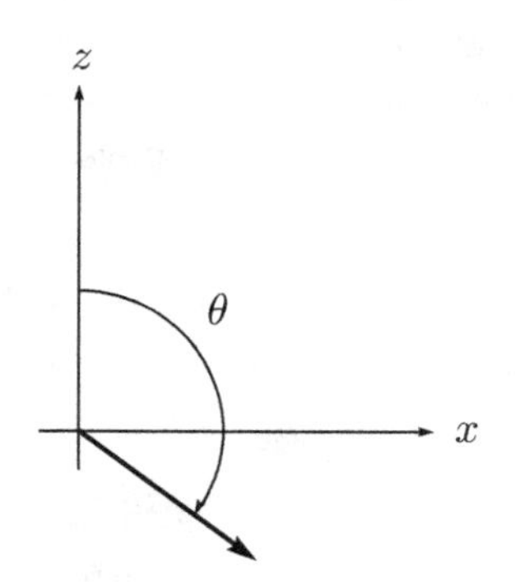

3.5 How can I specify a quantal state?

We introduced the Dirac notation for quantal states on page 90, but haven't yet fleshed out that notation by specifying a state mathematically. Start with an analogy:

3.5.1 *How can I specify a position vector?*

We are so used to writing down the position vector $\vec{r}$ that we rarely stop to ask ourselves what it means. But the plain fact is that whenever we measure a length (say, with a meter stick) we find not a vector, but a single number! Experiments measure never the vector $\vec{r}$ but always a scalar — the dot product between $\vec{r}$ and some other vector, call it $\vec{s}$ for "some other".

If we know the dot product between $\vec{r}$ and every vector $\vec{s}$, then we know everything there is to know about $\vec{r}$. Does this mean that to specify $\vec{r}$, we must keep a list of all possible dot products $\vec{s} \cdot \vec{r}$? Of course not... such a list would be infinitely long!

You know that if you write $\vec{r}$ in terms of an orthonormal basis $\{\hat{i}, \hat{j}, \hat{k}\}$, namely

$$\vec{r} = r_x\hat{i} + r_y\hat{j} + r_z\hat{k} \tag{3.25}$$

where $r_x = \hat{i} \cdot \vec{r}$, $r_y = \hat{j} \cdot \vec{r}$, and $r_z = \hat{k} \cdot \vec{r}$, then you've specified the vector. Why? Because if you know the triplet (r_x, r_y, r_z) and the triplet (s_x, s_y, s_z), then you can easily find the desired dot product

$$\vec{s} \cdot \vec{r} = \begin{pmatrix} s_x & s_y & s_z \end{pmatrix} \begin{pmatrix} r_x \\ r_y \\ r_z \end{pmatrix} = s_x r_x + s_y r_y + s_z r_z. \tag{3.26}$$

It's a lot more compact to specify the vector through three dot products — namely $\hat{i} \cdot \vec{r}$, $\hat{j} \cdot \vec{r}$, and $\hat{k} \cdot \vec{r}$ — from which you can readily calculate an infinite number of desired dot products, than it is to list all infinity dot products themselves!

3.5.2 *How can I specify a quantal state?*

Like the position vector $\vec{r}$, the quantal state $|\psi\rangle$ cannot by itself be measured. But if we determine (through some combination of analyzer experiments, interference experiments, and convention) the amplitude $\langle\sigma|\psi\rangle$ for every possible state $|\sigma\rangle$, then we know everything there is to know about $|\psi\rangle$. Is there some compact way of specifying the state, or do we have to keep an infinitely long list of all these amplitudes?

This nut is cracked through the interference experiment result

$$\langle\sigma|\psi\rangle = \langle\sigma|\theta+\rangle\langle\theta+|\psi\rangle + \langle\sigma|\theta-\rangle\langle\theta-|\psi\rangle, \tag{3.27}$$

which simply says, in symbols, that the atom exits an interferometer in the same state in which it entered (see equation 3.11). It gets hard to keep track of all these symbols, so I'll introduce the names

$$\begin{aligned}\langle\theta+|\psi\rangle &= \psi_+\\ \langle\theta-|\psi\rangle &= \psi_-\end{aligned}$$

and

$$\begin{aligned}\langle\theta+|\sigma\rangle &= \sigma_+\\ \langle\theta-|\sigma\rangle &= \sigma_-.\end{aligned}$$

From the reversal-conjugation relation, this means

$$\begin{aligned}\langle\sigma|\theta+\rangle &= \sigma_+^*\\ \langle\sigma|\theta-\rangle &= \sigma_-^*.\end{aligned}$$

In terms of these symbols, the interference result (3.27) is

$$\langle\sigma|\psi\rangle = \sigma_+^*\psi_+ + \sigma_-^*\psi_- = \begin{pmatrix}\sigma_+^* & \sigma_-^*\end{pmatrix}\begin{pmatrix}\psi_+\\ \psi_-\end{pmatrix}. \tag{3.28}$$

And this is our shortcut! By keeping track of only two amplitudes, ψ_+ and ψ_-, for each state, we can readily calculate any amplitude desired. We don't have to keep an infinitely long list of amplitudes.

This dot product result for computing amplitude is so useful and so convenient that sometimes people say the amplitude *is* a dot product. No. The amplitude reflects analyzer experiments, plus interference experiments, plus convention. The dot product is a powerful mathematical tool for computing amplitudes. (A parallel situation: There are many ways to find the latitude and longitude coordinates for a point on the Earth's surface, but the easiest is to use a GPS device. Some people are so enamored of this ease that they call the latitude and longitude the "GPS coordinates". But in fact the coordinates were established long before the Global Positioning System was built.)

3.5.3 *What is a basis?*

For vectors in three-dimensional space, an orthonormal basis such as $\{\hat{\imath}, \hat{\jmath}, \hat{k}\}$ is a set of three vectors of unit magnitude perpendicular to each other. As we've seen, the importance of a basis is that every vector $\vec{r}$ can be represented as a sum over these basis vectors,

$$\vec{r} = r_x\hat{\imath} + r_y\hat{\jmath} + r_z\hat{k},$$

and hence any vector $\vec{r}$ can be conveniently represented through the triplet

$$\begin{pmatrix} r_x \\ r_y \\ r_z \end{pmatrix} = \begin{pmatrix} \hat{\imath}\cdot\vec{r} \\ \hat{\jmath}\cdot\vec{r} \\ \hat{k}\cdot\vec{r} \end{pmatrix}.$$

For quantal states, we've seen that a set of two states such as $\{|\theta+\rangle, |\theta-\rangle\}$ plays a similar role, so it too is called a basis. For the magnetic moment of a silver atom, two states $|a\rangle$ and $|b\rangle$ constitute a basis whenever $\langle a|b\rangle = 0$, and the analyzer experiment of section 2.1.3 shows that the states $|\theta+\rangle$ and $|\theta-\rangle$ certainly satisfy this requirement. In the basis $\{|a\rangle, |b\rangle\}$ an arbitrary state $|\psi\rangle$ can be conveniently represented through the pair of amplitudes

$$\begin{pmatrix} \langle a|\psi\rangle \\ \langle b|\psi\rangle \end{pmatrix}.$$

3.5.4 *Hilbert space*

We have learned to express a physical state as a mathematical entity — namely, using the $\{|a\rangle, |b\rangle\}$ basis, the state $|\psi\rangle$ is represented as a column matrix of amplitudes

$$\begin{pmatrix} \langle a|\psi\rangle \\ \langle b|\psi\rangle \end{pmatrix}.$$

This mathematical entity is called a "state vector in Hilbert[6] space".

For example, in the basis $\{|z+\rangle, |z-\rangle\}$ the state $|\theta+\rangle$ is represented by

$$\begin{pmatrix} \langle z+|\theta+\rangle \\ \langle z-|\theta+\rangle \end{pmatrix} = \begin{pmatrix} \cos(\theta/2) \\ \sin(\theta/2) \end{pmatrix}. \tag{3.29}$$

[6]The German mathematician David Hilbert (1862–1943) made contributions to functional analysis, geometry, mathematical physics, and other areas. He formalized and extended the concept of a vector space. Hilbert and Albert Einstein raced to uncover the field equations of general relativity, but Einstein beat Hilbert by a matter of weeks.

Whereas (in light of equation 3.23) in the basis $\{|x+\rangle, |x-\rangle\}$ that same state $|\theta+\rangle$ is represented by the different column matrix

$$\begin{pmatrix} \langle x+|\theta+\rangle \\ \langle x-|\theta+\rangle \end{pmatrix} = \begin{pmatrix} \frac{1}{\sqrt{2}}[\cos(\theta/2) + \sin(\theta/2)] \\ -\frac{1}{\sqrt{2}}[\cos(\theta/2) - \sin(\theta/2)] \end{pmatrix}. \tag{3.30}$$

Write down the interference experiment result twice

$$\begin{aligned} \langle a|\psi\rangle &= \langle a|z+\rangle\langle z+|\psi\rangle + \langle a|z-\rangle\langle z-|\psi\rangle \\ \langle b|\psi\rangle &= \langle b|z+\rangle\langle z+|\psi\rangle + \langle b|z-\rangle\langle z-|\psi\rangle \end{aligned}$$

and then write these two equations as one using column matrix notation

$$\begin{pmatrix} \langle a|\psi\rangle \\ \langle b|\psi\rangle \end{pmatrix} = \begin{pmatrix} \langle a|z+\rangle \\ \langle b|z+\rangle \end{pmatrix} \langle z+|\psi\rangle + \begin{pmatrix} \langle a|z-\rangle \\ \langle b|z-\rangle \end{pmatrix} \langle z-|\psi\rangle.$$

Notice the column matrix representations of states $|\psi\rangle$, $|z+\rangle$, and $|z-\rangle$, and write this equation as

$$|\psi\rangle = |z+\rangle\langle z+|\psi\rangle + |z-\rangle\langle z-|\psi\rangle. \tag{3.31}$$

And now we have a new thing under the sun. We never talk about adding together two classical states, nor multiplying them by numbers, but this equation gives us the meaning of such state addition in quantum mechanics. This is a new mathematical tool, it deserves a new name, and that name is "superposition". Superposition is the mathematical reflection of the physical phenomenon of interference, as in the sentence: "When an atom ambivates through an interferometer, its state is a superposition of the state of an atom taking path a and the state of an atom taking path b."

Superposition is not familiar from daily life or from classical mechanics, but there is a story[7] that increases understanding: "A medieval European traveler returns home from a journey to India, and describes a rhinoceros as a sort of cross between a dragon and a unicorn." In this story the rhinoceros, an animal that is *not* familiar but that *does* exist, is described as intermediate (a "sort of cross") between two fantasy animals (the dragon and the unicorn) that *are* familiar (to the medieval European) but that *do not* exist.

Similarly, an atom in state $|z+\rangle$ ambivates through both paths of a horizontal interferometer. This action is *not* familiar but *does* happen, and it is characterized as a superposition (a "sort of cross") between two actions

[7]Invented by John D. Roberts, but first published in Robert T. Morrison and Robert N. Boyd, *Organic Chemistry*, second edition (Allyn & Bacon, Boston, 1966) page 318.

("taking path a" and "taking path b") that *are* familiar (to all of us steeped in the classical approximation) but that *do not* happen.

In principle, any calculation performed using the Hilbert space representation of states could be performed by considering suitable, cleverly designed analyzer and interference experiments. But it's a lot easier to use the abstract Hilbert space machinery. (Similarly, any result in electrostatics could be found using Coulomb's Law, but it's a lot easier to use the abstract electric field and electric potential. Any calculation involving vectors could be performed graphically, but it's a lot easier to use abstract components. Any addition or subtraction of whole numbers could be performed by counting out marbles, but it's a lot easier to use abstract mathematical tools like carrying and borrowing.)

3.5.5 *Peculiarities of state vectors*

Because state vectors are built from amplitudes, and amplitudes have peculiarities (see pages 95 and 101), it is natural that state vectors have similar peculiarities. For example, since the angle θ is the same as the angle $\theta + 360°$, I would expect that the state vector $|\theta+\rangle$ would be the same as the state vector $|(\theta + 360°)+\rangle$.

But in fact, in the $\{|z+\rangle, |z-\rangle\}$ basis, the state $|\theta+\rangle$ is represented by

$$\begin{pmatrix} \langle z+|\theta+\rangle \\ \langle z-|\theta+\rangle \end{pmatrix} = \begin{pmatrix} \cos(\theta/2) \\ \sin(\theta/2) \end{pmatrix}, \tag{3.32}$$

so the state $|(\theta + 360°)+\rangle$ is represented by

$$\begin{aligned} \begin{pmatrix} \langle z+|(\theta + 360°)+\rangle \\ \langle z-|(\theta + 360°)+\rangle \end{pmatrix} &= \begin{pmatrix} \cos((\theta + 360°)/2) \\ \sin((\theta + 360°)/2) \end{pmatrix} \\ &= \begin{pmatrix} \cos(\theta/2 + 180°) \\ \sin(\theta/2 + 180°) \end{pmatrix} = \begin{pmatrix} -\cos(\theta/2) \\ -\sin(\theta/2) \end{pmatrix}. \end{aligned} \tag{3.33}$$

So in fact $|\theta+\rangle = -|(\theta + 360°)+\rangle$. Bizarre!

This bizarreness is one facet of a general rule: If you multiply any state vector by a complex number with magnitude unity — a number such as -1, or i, or $\frac{1}{\sqrt{2}}(-1 + i)$, or $e^{2.7i}$ — a so-called "complex unit" or "phase factor" — then you get a different state vector that represents the same state. This fact is called "global phase freedom" — you are free to set the overall phase of your state vector for your own convenience. This general rule applies only for multiplying both elements of the state vector by the same complex unit: if you multiply the two elements with *different* complex units, you *will* obtain a vector representing a different state (see problem 3.8 on page 110).

3.5.6 *Names for position vectors*

The vector $\vec{r}$ is specified in the basis $\{\hat{\imath}, \hat{\jmath}, \hat{k}\}$ by the three components

$$\begin{pmatrix} r_x \\ r_y \\ r_z \end{pmatrix} = \begin{pmatrix} \hat{\imath} \cdot \vec{r} \\ \hat{\jmath} \cdot \vec{r} \\ \hat{k} \cdot \vec{r} \end{pmatrix}.$$

Because this component specification is so convenient, it is sometimes said that the vector $\vec{r}$ is not just specified, but is *equal* to this triplet of numbers. That's false.

Think of the vector $\vec{r} = 5\hat{\imath} + 5\hat{\jmath}$. It is represented in the basis $\{\hat{\imath}, \hat{\jmath}, \hat{k}\}$ by the triplet $(5, 5, 0)$. But this is not the only basis that exists. In the basis $\{\hat{\imath}' = (\hat{\imath}+\hat{\jmath})/\sqrt{2}, \hat{\jmath}' = (-\hat{\imath}+\hat{\jmath})/\sqrt{2}, \hat{k}\}$, that same vector is represented by the triplet $(5\sqrt{2}, 0, 0)$. If we had said that $\vec{r} = (5, 5, 0)$ and that $\vec{r} = (5\sqrt{2}, 0, 0)$, then we would be forced to conclude that $5 = 5\sqrt{2}$ and that $5 = 0$!

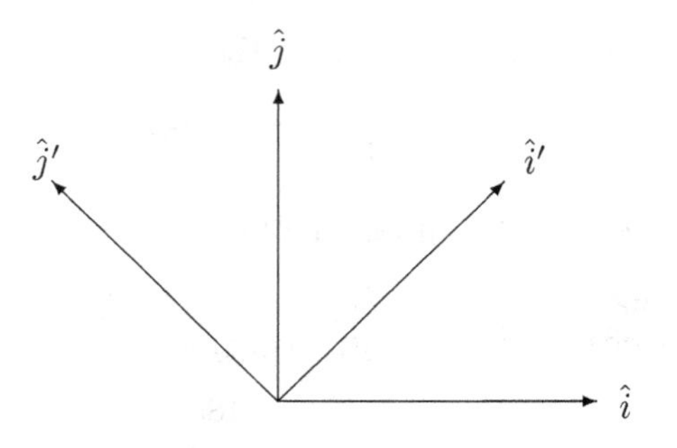

To specify a position vector $\vec{r}$, we use the components of $\vec{r}$ in a particular basis, usually denoted (r_x, r_y, r_z). We often write "$\vec{r} = (r_x, r_y, r_z)$" but in fact that's not exactly correct. The vector $\vec{r}$ represents a position — it is independent of basis. The row matrix (r_x, r_y, r_z) represents the components of that position vector in a particular basis — it is the "name" of the position in a particular basis. Instead of using an equals sign $=$ we use the symbol $\doteq$ to mean "represented by in a particular basis", as in "$\vec{r} \doteq$

$(5, 5, 0)$" meaning "the vector $\vec{r} = 5\hat{\imath} + 5\hat{\jmath}$ is represented by the triplet $(5, 5, 0)$ in the basis $\{\hat{\imath}, \hat{\jmath}, \hat{k}\}$".

Vectors are physical things: a caveman throwing a spear at a mammoth was performing addition of position vectors, even though the caveman didn't understand basis vectors or Cartesian coordinates. The concept of "position" was known to cavemen who did not have any concept of "basis".

3.5.7 *Names for quantal states*

We've been specifying a state like $|\psi\rangle = |17°+\rangle$ by stating the axis upon which the projection of $\vec{\mu}$ is definite and equal to $+\mu_B$ — in this case, the axis tilted 17° from the vertical.

Another way to specify a state $|\psi\rangle$ would be to give the amplitude that $|\psi\rangle$ is in any possible state: that is, to list $\langle\theta+|\psi\rangle$ and $\langle\theta-|\psi\rangle$ for all values of θ: $0° \leq \theta < 360°$. One of those amplitudes (in this case $\langle 17°+|\psi\rangle$) will have value 1, and finding this one amplitude would give us back the information in the specification $|17°+\rangle$. In some ways this is a more convenient specification because we don't have to look up amplitudes: they're right there in the list. On the other hand it is an awful lot of information to have to carry around.

The Hilbert space approach is a third way to specify a state that combines the brevity of the first way with the convenience of the second way. Instead of listing the amplitude $\langle\sigma|\psi\rangle$ for *every* state $|\sigma\rangle$ we list only the two amplitudes $\langle a|\psi\rangle$ and $\langle b|\phi\rangle$ for the elements $\{|a\rangle, |b\rangle\}$ of a basis. We've already seen (equation 3.28) how quantal interference then allows us to readily calculate any amplitude.

Just as we said "the position vector $\vec{r}$ is represented in the basis $\{\hat{\imath}, \hat{\jmath}, \hat{k}\}$ as $(1, 1, 0)$" or

$$\vec{r} \doteq (1, 1, 0),$$

so we say "the quantal state $|\psi\rangle$ is represented in the basis $\{|z+\rangle, |z-\rangle\}$ as

$$|\psi\rangle \doteq \begin{pmatrix} \langle z+|\psi\rangle \\ \langle z-|\psi\rangle \end{pmatrix}."$$

Problems

3.7 **Superposition and interference** (recommended problem)
On page 106 I wrote that "When an atom ambivates through an interferometer, its state is a superposition of the state of an atom taking path a and the state of an atom taking path b."

a. Write down a superposition equation reflecting this sentence for the interference experiment sketched on page 91.
b. Do the same for the interference experiment sketched on page 94.

3.8 **Representations** (recommended problem)
In the $\{|z+\rangle, |z-\rangle\}$ basis the state $|\psi\rangle$ is represented by

$$\begin{pmatrix} \psi_+ \\ \psi_- \end{pmatrix}.$$

(In other words, $\psi_+ = \langle z+|\psi\rangle$ and $\psi_- = \langle z-|\psi\rangle$.)

a. If ψ_+ and ψ_- are both real, show that there is one and only one axis upon which the projection of $\vec{\mu}$ has a definite, positive value, and find the angle between that axis and the z axis in terms of ψ_+ and ψ_-.
b. What would change if you multiplied both ψ_+ and ψ_- by the same phase factor (complex unit)?
c. What would change if you multiplied ψ_+ and ψ_- by different phase factors?

This problem invites the question "What if the ratio of ψ_+/ψ_- is not pure real?" When you study more quantum mechanics, you will find that in this case the axis upon which the projection of $\vec{\mu}$ has a definite, positive value is not in the x-z plane, but instead has a component in the y direction as well.

3.9 **Addition of states**
Some students in your class wonder "What does it mean to 'add two quantal states'? You never add two classical states." For the *Underground Guide to Quantum Mechanics* (see page 20) you decide to write four sentences interpreting the equation

$$|\psi\rangle = a|z+\rangle + b|z-\rangle \tag{3.34}$$

describing why you *can* add quantal states but *can't* add classical states. Your four sentences should include a formula for the amplitude a in terms of the states $|\psi\rangle$ and $|z+\rangle$.

3.10 **Names of six states, in two bases**

Write down the representations (the "names") of the states $|z+\rangle$, $|z-\rangle$, $|x+\rangle$, $|x-\rangle$, $|\theta+\rangle$, and $|\theta-\rangle$ in (a) the basis $\{|z+\rangle, |z-\rangle\}$ and in (b) the basis $\{|x+\rangle, |x-\rangle\}$.

3.11 **More peculiarities of states**

Because a vector pointing down at angle θ is the same as a vector pointing up at angle $\theta - 180°$, I would expect that $|\theta-\rangle = |(\theta - 180°)+\rangle$. Show that this expectation is false by uncovering the true relation between these two state vectors.

3.12 **Translation matrix**

(This problem requires background knowledge in the mathematics of matrix multiplication.)

Suppose that the representation of $|\psi\rangle$ in the basis $\{|z+\rangle, |z-\rangle\}$ is

$$\begin{pmatrix} \psi_+ \\ \psi_- \end{pmatrix} = \begin{pmatrix} \langle z+|\psi\rangle \\ \langle z-|\psi\rangle \end{pmatrix}.$$

The representation of $|\psi\rangle$ in the basis $\{|\theta+\rangle, |\theta-\rangle\}$ is just as good, and we call it

$$\begin{pmatrix} \psi'_+ \\ \psi'_- \end{pmatrix} = \begin{pmatrix} \langle \theta+|\psi\rangle \\ \langle \theta-|\psi\rangle \end{pmatrix}.$$

Show that you can "translate" between these two representations using the matrix multiplication

$$\begin{pmatrix} \psi'_+ \\ \psi'_- \end{pmatrix} = \begin{pmatrix} \cos(\theta/2) & \sin(\theta/2) \\ -\sin(\theta/2) & \cos(\theta/2) \end{pmatrix} \begin{pmatrix} \psi_+ \\ \psi_- \end{pmatrix}.$$

3.6 States for entangled systems

In the Einstein-Podolsky-Rosen experiment **(1)** on page 74, with two vertical analyzers, the initial state is represented by $|\psi\rangle$, and various possible final states are represented by $|\uparrow\downarrow\rangle$ and so forth, as shown below. (In this section all analyzers will be vertical, so we adopt the oft-used convention that writes $|z+\rangle$ as $|\uparrow\rangle$ and $|z-\rangle$ as $|\downarrow\rangle$.)

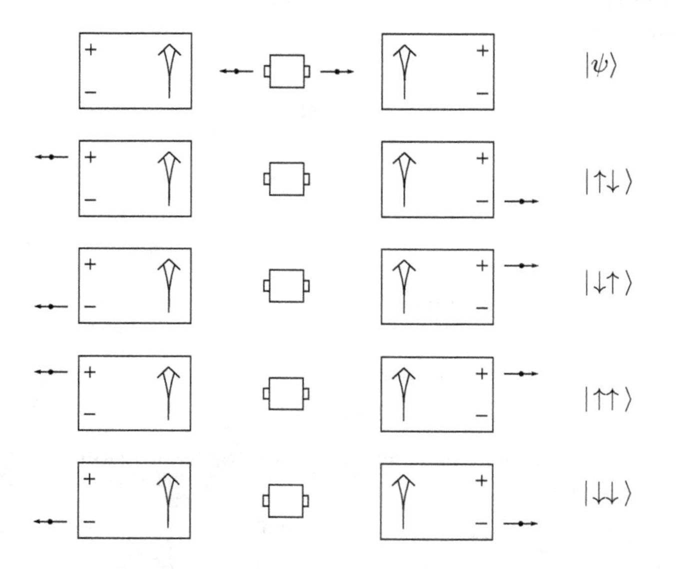

The experimental results tell us that

$$\begin{aligned}|\langle\uparrow\downarrow|\psi\rangle|^2 &= \tfrac{1}{2}\\ |\langle\downarrow\uparrow|\psi\rangle|^2 &= \tfrac{1}{2}\\ |\langle\uparrow\uparrow|\psi\rangle|^2 &= 0\\ |\langle\downarrow\downarrow|\psi\rangle|^2 &= 0.\end{aligned} \tag{3.35}$$

Additional analysis (sketched in problem 6.8, "Normalization of singlet spin state") is needed to assign phases to these amplitudes. The results are

$$\begin{aligned}\langle\uparrow\downarrow|\psi\rangle &= +\tfrac{1}{\sqrt{2}}\\ \langle\downarrow\uparrow|\psi\rangle &= -\tfrac{1}{\sqrt{2}}\\ \langle\uparrow\uparrow|\psi\rangle &= 0\\ \langle\downarrow\downarrow|\psi\rangle &= 0.\end{aligned} \tag{3.36}$$

Using the generalization of equation (3.31) for a four-state basis, these results tell us that

$$\begin{aligned}|\psi\rangle &= |\uparrow\downarrow\rangle\langle\uparrow\downarrow|\psi\rangle + |\downarrow\uparrow\rangle\langle\downarrow\uparrow|\psi\rangle + |\uparrow\uparrow\rangle\langle\uparrow\uparrow|\psi\rangle + |\downarrow\downarrow\rangle\langle\downarrow\downarrow|\psi\rangle \\ &= \tfrac{1}{\sqrt{2}}(|\uparrow\downarrow\rangle - |\downarrow\uparrow\rangle).\end{aligned} \tag{3.37}$$

A simple derivation, with profound implications.

3.6.1 *State pertains to system, not to atom*

In this entangled situation there is no such thing as an "amplitude for the right atom to exit from the + port," because the probability for the right atom to exit from the + port depends on whether the left atom exits the + or the − port. The pair of atoms has a state, but the right atom by itself *doesn't have a state*, in the same way that an atom passing through an interferometer *doesn't have a position* and that love *doesn't have a color.*

Leonard Susskind[8] puts it this way: If entangled states existed in auto mechanics as well as quantum mechanics, then an auto mechanic might tell you "I know everything about your car but ... I can't tell you anything about any of its parts."

3.6.2 *"Collapse of the state vector"*

Set up this EPR experiment with the left analyzer 100 kilometers from the source, and the right analyzer 101 kilometers from the source. As soon as the left atom comes out of its − port, then it is known that the right atom will come out if its + port. The system is no longer in the entangled state $\frac{1}{\sqrt{2}}(|\uparrow\downarrow\rangle - |\downarrow\uparrow\rangle)$; instead the left atom is in state $|\downarrow\rangle$ and the right atom is in state $|\uparrow\rangle$. The state of the right atom has changed (some say it has "collapsed") despite the fact that it is 200 kilometers from the left analyzer that did the state changing!

This fact disturbs those who hold the misconception that states are physical things located out in space like nitrogen molecules, because it seems that information about state has made an instantaneous jump across 200 kilometers. In fact no information has been transferred from left to right: true, Alice at the left interferometer knows that the right atom will

[8]Leonard Susskind and Art Friedman, *Quantum Mechanics: The Theoretical Minimum* (Basic Books, New York, 2014) page xii.

exit the + port 201 kilometers away, but Bob at the right interferometer *doesn't* have this information and won't unless she tells him in some conventional, light-speed-or-slower fashion.[9]

If Alice could in some magical way manipulate her atom to ensure that it would exit the − port, then she *could* send a message instantaneously. But Alice does not possess magic, so she cannot manipulate the left-bound atom in this way. Neither Alice, nor Bob, *nor even the left-bound atom itself* knows from which port it will exit. Neither Alice, nor Bob, *nor even the left-bound atom itself* can influence from which port it will exit.

3.6.3 *Measurement and entanglement*

Back in section 2.4, "Light on the atoms" (page 70), we discussed the character of "observation" or "measurment" in quantum mechanics. Let's bring our new machinery concerning quantal states to bear on this situation.

The figure on the next page shows, in the top panel, a potential measurement about to happen. An atom (represented by a black dot) in state $|z+\rangle$ approaches a horizontal interferometer at the same time that a photon (represented by a white dot) approaches path a of that interferometer.

We employ a simplified model in which the photon either misses the atom, in which case it continues undeflected upward, or else the photon interacts with the atom, in which case it is deflected outward from the page. In this model there are four possible outcomes, shown in the bottom four panels of the figure.

After this potential measurement, the system of photon plus atom is in an entangled state: the states shown on the right must list *both* the condition of the photon ("up" or "out") *and* the condition of the atom (+ or −).

If the photon misses the atom, then the atom must emerge from the + port of the analyzer: there is zero probability that the system has final state $|\text{up}; -\rangle$. But if the photon interacts with the atom, then the atom might emerge from either port: there is non-zero probability that the system has

[9] If you are familiar with gauges in electrodynamics, you will find quantal state similar to the Coulomb gauge. In the Coulomb gauge, the electric potential at a point in space changes the instant that any charged particle moves, regardless of how far away that charged particle is. This does not imply that information moves instantly, because electric potential by itself is not measurable. The same applies for quantal state.

final state $|\text{out}; -\rangle$. These two states are exactly the same as far as the atom is concerned; they differ only in the position of the photon.

If we focus only on the atom, we would say that something strange has happened (a "measurement" at path a) that enabled the atom to emerge from the $-$ port which (in the absence of "measurement") that atom would never do. But if we focus on the entire system of photon plus atom, then it is an issue of entanglement, not of measurement.

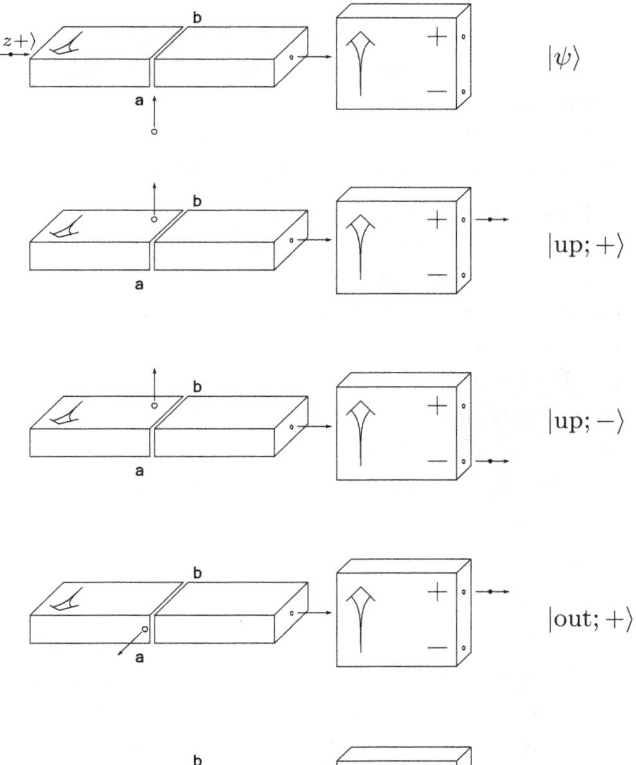

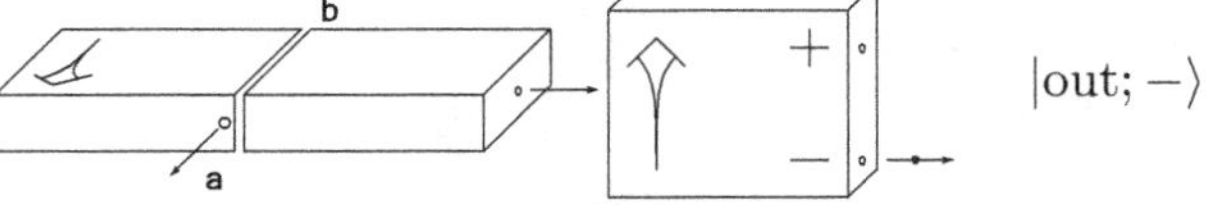

Problem

3.13 **Amplitudes for "Measurement and entanglement"**
Suppose that, in the "simplified model" for measurement and entanglement, the probability for photon deflection is $\frac{1}{5}$. Find the four probabilities $|\langle \text{up}; +|\psi\rangle|^2$, $|\langle \text{up}; -|\psi\rangle|^2$, $|\langle \text{out}; +|\psi\rangle|^2$, and $|\langle \text{out}; -|\psi\rangle|^2$.

3.7 Are states "real"?

This is a philosophical question for which there is no specific meaning and hence no specific answer. But in my opinion, states are mathematical tools that enable us to efficiently and accurately calculate the probabilities that can be found through repeated analyzer experiments, interference experiments, and indeed all experiments. They are not physically "real".

Indeed, it is possible to formulate quantum mechanics in such a way that probabilities and amplitudes are found without using the mathematical tool of "state" at all. Richard Feynman and Albert Hibbs do just this in their 1965 book *Quantum Mechanics and Path Integrals.* States do not make an appearance until deep into their book, and even when they do appear they are not essential. The Feynman "sum over histories" formulation described in that book is, for me, the most intuitively appealing approach to quantum mechanics. There is, however, a price to be paid for this appeal: it's very difficult to work problems in the Feynman formulation.

3.8 What is a qubit?

At the end of the last chapter (on page 88) we listed several so-called "two-state systems" or "spin-$\frac{1}{2}$ systems" or "qubit systems". You might have found these terms strange: There are an infinite number of states for the magnetic moment of a silver atom: $|z+\rangle$, $|1^\circ+\rangle$, $|2^\circ+\rangle$, and so forth. Where does the name "two-state system" come from? You now see the answer: it's short for "two-basis-state system".

The term "spin" originated in the 1920s when it was thought that an electron was a classical charged rigid sphere that created a magnetic moment through spinning about an axis. A residual of that history is that people still call[10] the state $|z+\rangle$ by the name "spin up" and by the symbol $|\uparrow\rangle$, and the state $|z-\rangle$ by "spin down" and $|\downarrow\rangle$. (Sometimes the association is made in the opposite way.) Meanwhile the state $|x+\rangle$ is given the name "spin sideways" and the symbol $|\rightarrow\rangle$.

Today, two-basis-state systems are more often called "qubit" systems from the term used in quantum information processing. In a classical computer, like the ones we use today, a bit of information can be represented physically by a patch of magnetic material on a disk: the patch magnetized "up" is interpreted as a 1, the patch magnetized "down" is interpreted as a 0. Those are the only two possibilities. In a quantum computer, a qubit of information can be represented physically by the magnetic moment of a silver atom: the atom in state $|z+\rangle$ is interpreted as $|1\rangle$, the atom in state $|z-\rangle$ is interpreted as $|0\rangle$. But the atom might be in any (normalized) superposition $a|1\rangle + b|0\rangle$, so rather than two possibilities there are an infinite number.

Furthermore, qubits can interfere with and become entangled with other qubits, options that are simply unavailable to classical bits. With more states, and more ways to interact, quantum computers can only be faster than classical computers, and even as I write these possibilities are being explored.

In today's state of technology, quantum computers are hard to build, and they may never live up to their promise. But maybe they will.

Chapters 2 and 3 have focused on two-basis-state systems, but of course nature provides other systems as well. For example, the magnetic moment of a nitrogen atom (mentioned on page 45) is a "four-basis-state" system, where one basis is

$$|z;+2\rangle, \quad |z;+1\rangle, \quad |z;-1\rangle, \quad |z;-2\rangle. \tag{3.38}$$

In fact, the next chapter shifts our focus to a system with an infinite number of basis states.

[10]The very most precise and pedantic people restrict the term "spin" to elementary particles, such as electrons and neutrinos. For composite systems like the silver atom they speak instead of "the total angular momentum $\vec{J}$ of the silver atom in its ground state, projected on a given axis, and divided by $\hbar$." For me, the payoff in precision is not worth the penalty in polysyllables.

Problem

3.14 **Questions** (recommended problem)
Update your list of quantum mechanics questions that you started at problem 1.17 on page 46. Write down new questions and, if you have uncovered answers to any of your old questions, write them down briefly.

⟦For example, one of my questions would be: "I'd like to see a proof that the global phase freedom mentioned on page 107, which obviously changes the amplitudes computed, does not change any experimentally accessible result."⟧

Chapter 4

The Quantum Mechanics of Position

In the last two chapters we've studied the quantum mechanics of a silver atom's magnetic moment, and we got a lot out of it: we uncovered the phenomena of quantization and interference and entanglement; we found how to use amplitude as a mathematical tool to predict probabilities; we learned about quantum mechanical states. If all of this makes you feel weak and dizzy, that's a good thing: Niels Bohr pointed out that "those who are not shocked when they first come across quantum theory cannot possibly have understood it."[1] As profitable as this has been, we knew from the start (page 47) that eventually we would need to treat the quantum mechanics of position. Now is the time.

Chapters 2 and 3 treated the atom's magnetic moment but (to the extent possible) ignored the atom's position. This chapter starts off with the opposite approach: it treats only position and ignores magnetic moment. Section 4.12, "Position plus spin", at the end of this chapter welds the two aspects together.

4.1 Probability and probability density: One particle in one dimension

A single particle ambivates in one dimension. You know the story of quantum mechanics: The particle doesn't have a position. Yet if we measure the position (say, by shining a lamp), then we *will* find that it has a single position. However, because the particle started out without a position,

[1] Recalled by Werner Heisenberg in *Physics and Beyond* (Harper and Row, New York, 1971) page 206.

there is no way to predict the position found beforehand: instead, quantum mechanics predicts probabilities.

But what exactly does this mean? There are an infinite number of points along any line, no matter how short. If there were a finite probability at each of these points, the total probability would be infinity. But the total probability must be one! To resolve this essential technical issue, we look at a different situation involving probability along a line.

You are studying the behavior of ants in an ant farm. (Ant farm: two panes of glass close together, with sand and ants and ant food between the panes.) The ant farm is 100.0 cm long. You paint one ant red, and 9741 times you look at the ant farm and measure (to the nearest millimeter) the distance of the red ant from the left edge of the farm.

You are left with 9741 raw numbers, and a conundrum: how should you present these numbers to help draw conclusions about ant behavior?

The best way is to conceptually divide the ant farm into bins. Start with five equal bins: locations from 0.0 cm to 20.0 cm are in the first bin, from 20.0 cm to 40.0 cm in the second, from 40.0 cm to 60.0 cm in the third, from 60.0 cm to 80.0 cm in the fourth, and from 80.0 cm to 100.0 cm in the fifth. Find the number of times the ant was in the first bin, and divide by the total number of observations (9741) to find the probability that the ant was in the first bin. Similarly for the other bins. You will produce a graph like this:

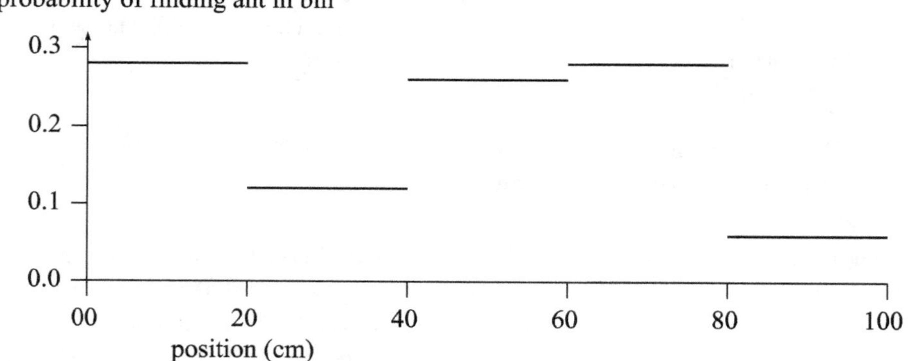

The five probabilities sum to 1, as they must.

Now you want more detail about the ant's location. Instead of dividing the ant farm into five conceptual bins each of width 20.0 cm, divide it into ten bins each of width 10.0 cm. The probabilities now look like:

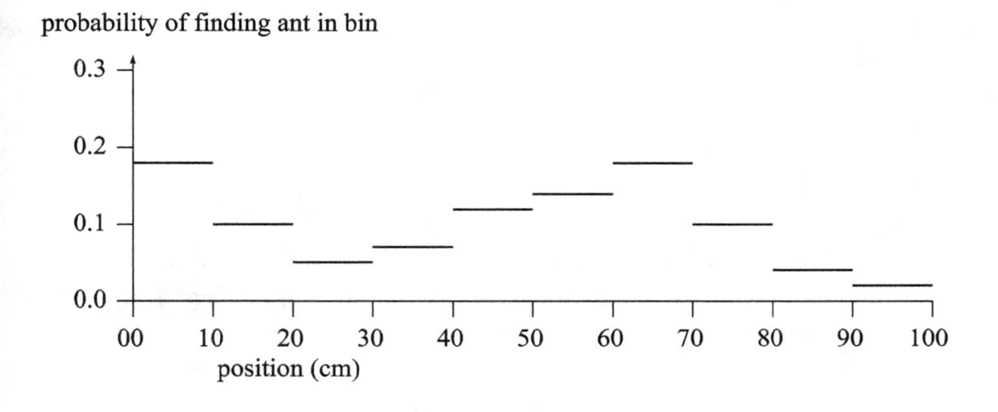

There are now ten probabilities, yet they still sum to 1, so the probabilities are each smaller. (For example, the first graph shows a probability of 0.28 for the ant appearing between 0.0 cm and 20.0 cm. The second graph shows probability 0.18 for the ant appearing between 0.0 cm and 10.0 cm and probability 0.10 for the ant appearing between 10.0 cm and 20.0 cm. Sure enough $0.28 = 0.18 + 0.10$.)

If you want still more detail, you can divide the ant farm into fifty bins, each of width 2.0 cm, as in:

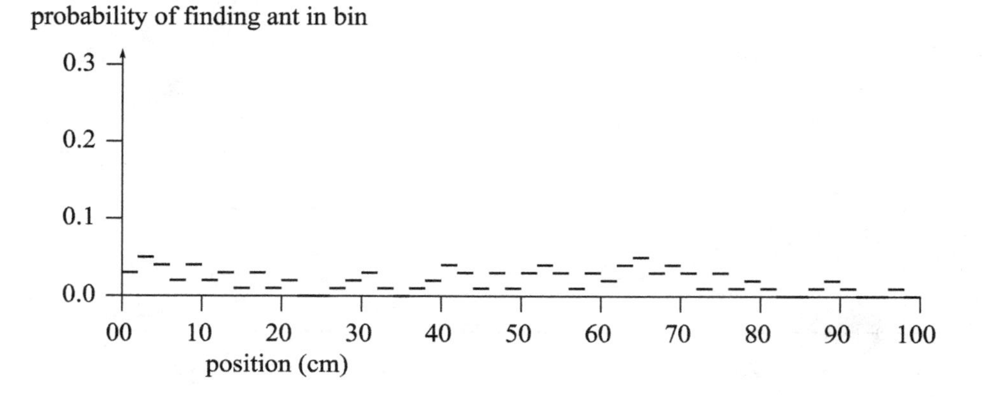

These fifty probabilities must still sum to 1, so the individual probabilities are smaller still.

You could continue this process, making the bins smaller and smaller, and every bin probability would approach zero. In symbols, if the probability for appearing in the bin surrounding point x_0 is called P_0, and the width of each bin is called Δx, we have that

$$\lim_{\Delta x \to 0} P_0 = 0$$

for all points x_0. This is true but provides no information whatsoever about ant behavior!

To get information, focus not on the bin probability but on the so-called *probability density*, defined as

$$\lim_{\Delta x \to 0} \frac{P_0}{\Delta x} \equiv \rho(x_0).$$

In terms of probability density, we say that "the probability of finding the ant in a small window of width w centered on x_0 is approximately $\rho(x_0)w$, and this approximation grows better and better as the window grows narrower and narrower." And that "the probability of finding the ant between x_A and x_B is

$$\int_{x_A}^{x_B} \rho(x)\, dx."$$

The fact that the bin probabilities, summed over all bins, is unity, or in symbols

$$\sum_i P_i = 1,$$

becomes, in the limit $\Delta x \to 0$,

$$\sum_i P_i \approx \sum_i \rho(x_i)\Delta x \to \int_{0.0\,\text{cm}}^{100.0\,\text{cm}} \rho(x)\,dx = 1.$$

This property of probability densities is called "normalization".

Problem

4.1 **Mean and standard deviation for an ant** (essential problem)
The mean[2] ant position $\langle x \rangle$ is given by summing all the position measurements and dividing that sum by the number of measurements (in this case 9741). Using P_i for the probability of the ant appearing in bin i and x_i for the position of the center of bin i, argue that this mean position is given approximately by

$$\sum_{\text{bin } i} x_i P_i$$

and that the approximation grows better and better as the bins grow narrower and narrower. The formula becomes exact when $\Delta x \to 0$, so show that the mean value is given by

$$\langle x \rangle = \int_{0.0\,\text{cm}}^{100.0\,\text{cm}} x\rho(x)\,dx. \tag{4.1}$$

In the same way, argue that the standard deviation of ant position is

$$\sqrt{\int_{0.0\,\text{cm}}^{100.0\,\text{cm}} (x - \langle x \rangle)^2 \rho(x)\,dx}. \tag{4.2}$$

[2]The "mean value" is also called the "average value" and sometimes the "expectation value". The latter name is particularly poor. If you toss a die, the mean value of the number facing up is 3.5. Yet no one expects to toss a die and find the number 3.5 facing up!

4.2 Probability amplitude

The probability considerations for one ant walking in one dimension are directly analogous to the probability considerations for one quantal particle ambivating in one dimension. A graph with five bins like the one on page 120 approximates the quantal particle as a five-state system. A graph with ten bins like the one on page 121 approximates the quantal particle as a ten-state system. A graph with fifty bins like the one on page 122 approximates the quantal particle as a fifty-state system. You know the drill of quantum mechanics: in all these cases the bin probability P_0 will be related to some sort of bin amplitude ψ_0 through $P_0 = |\psi_0|^2$. How does bin amplitude behave as $\Delta x \to 0$? Because

$$\frac{P_0}{\Delta x} = \frac{|\psi_0|^2}{\Delta x} \to \rho(x_0), \quad \text{we will have} \quad \frac{\psi_0}{\sqrt{\Delta x}} \to \psi(x_0),$$

where, for any point x_0, the probability density is

$$\rho(x_0) = |\psi(x_0)|^2. \tag{4.3}$$

What would be a good name for this function $\psi(x)$? I like the name "amplitude density". It's not really a density: a density would have dimensions 1/[length], whereas $\psi(x)$ has dimensions $1/\sqrt{[\text{length}]}$. But it's closer to a density than it is to anything else. Unfortunately, someone else (namely Schrödinger[3]) got to name it before I came up with this sensible name, and that name has stuck. It's called "wavefunction".

The normalization condition for wavefunction is

$$\int_{-\infty}^{+\infty} |\psi(x)|^2 \, dx = 1. \tag{4.4}$$

You should check for yourself that this equation is dimensionally consistent.

The global phase freedom described for qubit systems on page 107 applies for wavefunctions as well: If you multiply any wavefunction by a complex number with magnitude unity — called a "complex unit" or "phase

[3]Erwin Schrödinger (1887–1961) had interests in physics, biology, philosophy, and Eastern religion. Born in Vienna, he held physics faculty positions in Germany, Poland, and Switzerland. In 1926 he developed the concept of wavefunction and discovered the quantum mechanical time evolution equation (4.12) that now bears his name. This led, in 1927, to a prestigious appointment in Berlin. In 1933, disgusted with the Nazi regime, he left Berlin for Oxford, England. He held several positions in various cities before ending up in Dublin. There, in 1944, he wrote a book titled *What is Life?* which stimulated interest in what had previously been a backwater of science: biochemistry.

factor" — then you get a different wavefunction that represents the same state. You must multiply by a *number* with magnitude unity, not a *function* with magnitude unity: the wavefunctions $\psi(x)$ and $e^{i\delta}\psi(x)$ represent the same state, but the wavefunction $e^{i\delta(x)}\psi(x)$ represents a different state.

Keep in mind that to specify a quantal state we must know amplitudes (wavefunction) rather than merely probabilities (probability density). Just as in classical mechanics (see page 90) we say that we know a state when we know all the information needed to describe the system now and to predict its future. The probability density $\rho(x)$ alone tells you a lot about the state right now, but cannot predict how the state will change in the future. Knowing probability density alone in quantum mechanics is like knowing the position alone in classical mechanics: The probability density gives a lot of information about *now*, but not enough information to predict the future.

Problems

4.2 Mean and standard deviation for a quantal particle

(essential problem)

For any function $f(x)$, define the mean value

$$\langle f(x)\rangle = \int_{-\infty}^{+\infty} f(x)|\psi(x)|^2\,dx. \tag{4.5}$$

Show that the mean position is $\langle x\rangle$ and that the standard deviation of position Δx is given through

$$(\Delta x)^2 = \langle x^2\rangle - \langle x\rangle^2. \tag{4.6}$$

4.3 What we say makes no sense. What do we mean?

It sounds strange to say "The particle with wavefunction $\psi(x)$ doesn't have a position and its mean position is

$$\int_{-\infty}^{+\infty} x|\psi(x)|^2\,dx."$$

Write two or three sentences unpacking what this sentence really means.

4.4 Bump wavefunction

The parabolic bump wavefunction is defined as

$$\psi(x) = \begin{cases} 0 & x < 0 \\ -ax(x-L) & 0 \le x \le L \\ 0 & L < x \end{cases}.$$

Find the mean position and standard deviation of position.

4.3 How does wavefunction change with time?

I'm going to throw down three equations. First, the classical formula for energy,

$$E = \frac{p^2}{2m} + V, \tag{4.7}$$

where V is the potential energy. Second, the Einstein and de Broglie relations for energy and momentum (1.21) and (1.24)

$$E = \hbar\omega \quad \text{and} \quad p = \hbar k. \tag{4.8}$$

Third, the particular wavefunction

$$\psi(x,t) = Ae^{i(kx-\omega t)}. \tag{4.9}$$

Plugging equations (4.8) mindlessly into equation (4.7) we obtain

$$\hbar\omega = \frac{\hbar^2 k^2}{2m} + V$$

and multiplying both sides by $\psi(x,t)$ gives

$$\hbar\omega\psi(x,t) = \frac{\hbar^2 k^2}{2m}\psi(x,t) + V\psi(x,t). \tag{4.10}$$

Meanwhile, the particular wavefunction (4.9) satisfies

$$\omega\psi(x,t) = \frac{1}{-i}\frac{\partial\psi}{\partial t}; \quad k\psi(x,t) = \frac{1}{i}\frac{\partial\psi}{\partial x}; \quad k^2\psi(x,t) = \frac{1}{i^2}\frac{\partial^2\psi}{\partial x^2} = -\frac{\partial^2\psi}{\partial x^2}.$$

Plugging *these* into equation (4.10) gives

$$\hbar\left(\frac{1}{-i}\frac{\partial\psi}{\partial t}\right) = \frac{\hbar^2}{2m}\left(-\frac{\partial^2\psi}{\partial x^2}\right) + V\psi(x,t), \tag{4.11}$$

which rearranges to

$$\frac{\partial\psi(x,t)}{\partial t} = -\frac{i}{\hbar}\left[-\frac{\hbar^2}{2m}\frac{\partial^2\psi(x,t)}{\partial x^2} + V(x)\psi(x,t)\right]. \tag{4.12}$$

I concede from the very start that this is a stupid argument, and that if you had proposed it to me I would have gone ballistic. First, equation (4.7) is a classical fact plopped mindlessly into a quantal argument. Second, the Einstein relation (4.8) applies to photons, not to massive particles. Third, there are many possible wavefunctions other than equation (4.9). The unjustified change of potential energy value V in equation (4.11) to potential energy function $V(x)$ in equation (4.12) merely adds insult to injury. The only good thing I can say about this equation is that it's dimensionally consistent.

Oh, and one more thing. The equation is correct. Despite its dubious provenance, experimental tests have demonstrated to everyone's satisfaction that wavefunction *really does* evolve in time this way. (I must qualify: wavefunction evolves this way in a wide range of situations: non-relativistic, no magnetic field, no friction or any other non-conservative force, and where the particle's magnetic moment is unimportant.)

This equation for time evolution in quantal systems plays the same central role in quantum mechanics that $\vec{F} = m\vec{a}$ does in classical mechanics. And just as $\vec{F} = m\vec{a}$ cannot be derived, only motivated and then tested experimentally, so this time-evolution result cannot be derived. The motivation is lame, but the experimental tests are impressive and cannot be ignored.

This time evolution equation has a name: it is "the Schrödinger equation".

4.4 Wavefunction: Two particles in one or three dimensions

We will soon work on solving the Schrödinger equation for one particle in one dimension, but first we ask how to describe two particles in one dimension.

Two particles, say an electron and a neutron, ambivate in one dimension. As before, we start with a grid of bins in one-dimensional space:

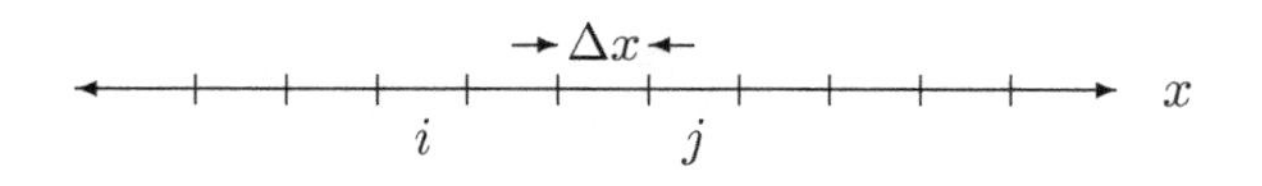

We ask for the probability that the electron will be found in bin i and the neutron will be found in bin j, and call the result $P_{i,j}$. Although our situation is one-dimensional, this question generates a two-dimensional array of probabilities.

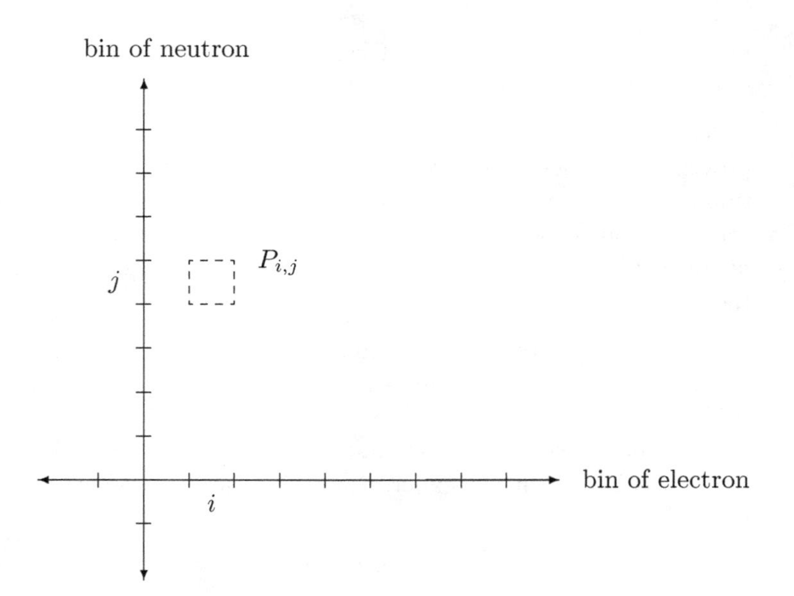

To produce a probability density, we must divide the bin probability $P_{i,j}$ by $(\Delta x)^2$, and then take the limit as $\Delta x \to 0$, resulting in

$$\frac{P_{i,j}}{(\Delta x)^2} \to \rho(x_e, x_n).$$

So the probability of finding an electron within a narrow window of width w centered on $x_e = 5$ and finding the neutron within a narrow window of width u centered on $x_n = 9$ is approximately $\rho(5,9)wu$, and this approximation grows better and better as the two windows grow narrower and narrower.

The bin amplitude is $\psi_{i,j}$ with $P_{i,j} = |\psi_{i,j}|^2$. To turn a bin amplitude into a wavefunction, divide by $\sqrt{(\Delta x)^2} = \Delta x$ and take the limit

$$\lim_{\Delta x \to 0} \frac{\psi_{i,j}}{\Delta x} = \psi(x_e, x_n). \tag{4.13}$$

This wavefunction has dimensions 1/[length].

The generalization to more particles and higher dimensionality is straightforward. For a single electron in three-dimensional space, the wavefunction $\psi(\vec{x})$ has dimensions $1/[\text{length}]^{3/2}$. For an electron and a neutron in three-dimensional space, the wavefunction $\psi(\vec{x}_e, \vec{x}_n)$ has dimensions $1/[\text{length}]^3$. Note carefully: For a two-particle system, the state is specified by one function $\psi(\vec{x}_e, \vec{x}_n)$ of six variables. It is *not* specified by two functions of three variables, with $\psi_e(\vec{x})$ giving the state of the electron and

$\psi_n(\vec{x})$ giving the state of the neutron. There are four consequences of this simple yet profound observation.

First, the wavefunction (like amplitude in general) is a mathematical tool for calculating the results of experiments; it is not physically "real". I have mentioned this before, but it particularly stands out here. Even for a system as simple as two particles, the wavefunction does not exist in ordinary three-dimensional space, but in the six-dimensional "configuration space", as it is called. I don't care how clever or talented an experimentalist you are: you cannot insert an instrument into six-dimensional space in order to measure wavefunction.

Second, wavefunction is associated with a *system*, not with a *particle*. If you're interested in a single electron and you say "the wavefunction of the electron", then you're technically incorrect — you should say "the wavefunction of the system consisting of a single electron" — but no one will go ballistic and say that you are in thrall of a deep misconception. However, if you're interested in a pair of particles (an electron and a neutron, for instance) and you say "the wavefunction of the electron", then someone (namely me) *will* go ballistic because you *are* in thrall of a deep misconception.

Third, it might happen that the wavefunction factorizes:

$$\psi(\vec{x}_e, \vec{x}_n) = \psi_e(\vec{x}_e)\psi_n(\vec{x}_n).$$

In this case the electron has state $\psi_e(\vec{x}_e)$ and the neutron has state $\psi_n(\vec{x}_n)$. Such a peculiar case is called "non-entangled". But in all other cases the state is called "entangled" and the individual particles making up the system *do not have states.* The system has a state, namely $\psi(\vec{x}_e, \vec{x}_n)$, but there is no state for the electron and no state for the neutron, in exactly the same sense that there is no position for a silver atom ambivating through an interferometer.

Fourth, quantum mechanics is intricate. To understand this point, contrast the description needed in classical versus quantum mechanics.

How does one describe the state of a single classical particle moving in one dimension? It requires two numbers: a position and a velocity. Two particles moving in one dimension require merely that we specify the state of each particle: four numbers. Similarly specifying the state of three particles require six numbers and N particles require $2N$ numbers. Exactly the same specification counts hold if the particle moves relativistically.

How, in contrast, does one describe the state of a single quantal particle ambivating in one dimension? Here an issue arises at the very start, because the specification is given through a complex-valued wavefunction $\psi(x)$. Technically the specification requires an infinite number of numbers! Let's approximate the wavefunction through its value on a grid of, say, 100 points. This suggests that a specification requires 200 real numbers, a complex number at each grid point, but global phase freedom means that we can always set one of those numbers to zero through an overall phase factor, and one number is not independent through the normalization requirement. The specification actually requires 198 independent real numbers.

How does one describe the state of two quantal particles ambivating in one dimension? Now the wavefunction is a function of two variables, $\psi(x_e, x_n)$. The wavefunction of the system is a function of two-dimensional configuration space, so an approximation of the accuracy established previously requires a 100×100 grid of points. Each grid point carries one complex number, and again overall phase and normalization reduce the number of real numbers required by two. For two particles the specification requires $2 \times (100)^2 - 2 = 19\,998$ independent real numbers.

Similarly, specifying the state of N quantal particles moving in one dimension requires a wavefunction in N-dimensional configuration space which (for a grid of the accuracy we've been using) is specified through $2 \times (100)^N - 2$ independent real numbers.

The specification of a quantal state not only requires more real numbers than the specification of the corresponding classical state, but that number increases exponentially rather than linearly with particle number N.

The fact that a quantal state holds more information than a classical state is the fundamental reason that a quantal computer can be (in principle) faster than a classical computer, and the basis for much of quantum information theory.

Relativity is different from classical physics, but no more complicated. Quantum mechanics, in contrast, is both *different* from and *richer* than classical physics. You may refer to this richness using terms like "splendor", or "abounding", or "intricate", or "ripe with possibilities". Or you may refer to it using terms like "complicated", or "messy", or "full of details likely to trip the innocent". It's your choice how to react to this richness, but you can't deny it.

4.5 Solving the Schrödinger time evolution equation for the infinite square well

Setup. A single particle is restricted to one dimension. In classical mechanics, the state of the particle is given through position and velocity: that is, we want to know the two functions of time

$$x(t); \quad v(t).$$

These functions stem from the solution to the ordinary differential equation (ODE) $\sum \vec{F} = m\vec{a}$, or, in this case,

$$\frac{d^2x(t)}{dt^2} = \frac{1}{m}F(x(t)) \tag{4.14}$$

subject to the given initial conditions

$$x(0) = x_0; \quad v(0) = v_0.$$

In quantum mechanics, the state of the particle is given through the wavefunction: that is, we want to know the two-variable function

$$\psi(x,t).$$

This is the solution of the Schrödinger partial differential equation (PDE)

$$\frac{\partial\psi(x,t)}{\partial t} = -\frac{i}{\hbar}\left[-\frac{\hbar^2}{2m}\frac{\partial^2\psi(x,t)}{\partial x^2} + V(x)\psi(x,t)\right], \tag{4.15}$$

subject to the given initial condition

$$\psi(x,0) = \psi_0(x).$$

⟦The classical time evolution equation (4.14) is *second* order in time, so there are two initial conditions: initial position and initial velocity. The quantal time evolution equation (4.15) is *first* order in time, so there is only one initial condition: initial wavefunction.⟧

Infinite square well. Since this is our first quantal time evolution problem, let's start out cautiously by choosing the easiest potential energy function: the so-called infinite square well[4] or "particle in a box":

$$V(x) = \begin{cases} \infty & \text{for } x \le 0 \\ 0 & \text{for } 0 < x < L \\ \infty & \text{for } L \le x \end{cases}$$

[4]Any potential energy function with a minimum is called a "well".

This is an approximate potential energy function for an electron added to a hydrocarbon chain molecule (a "conjugated polymer"), or for an atom trapped in a capped carbon nanotube.

The infinite square well is like the "perfectly rigid cylinder that rolls without slipping" in classical mechanics. It does not exactly exist in reality: any cylinder will be dented or cracked if hit hard enough. But it is a good model for some real situations. And it's certainly better to work with this model than it is to shrug your shoulders and say "I have no idea."

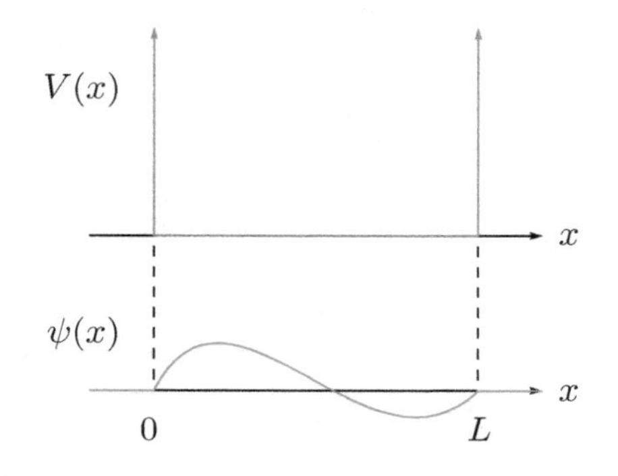

Figure 4.1: *The infinite square well potential energy function $V(x)$ in olive green, and a possible wavefunction $\psi(x)$ in red.*

It is reasonable (although not rigorously proven) that for the infinite square well

$$\psi(x,t) = \begin{cases} 0 & \text{for } x \leq 0 \\ \text{something} & \text{for } 0 < x < L \\ 0 & \text{for } L \leq x \end{cases}$$

and we adopt these conditions.

Strategy. The PDE is linear, so if we find some special solutions $f_1(x,t)$, $f_2(x,t)$, $f_3(x,t)$, ..., then we can generate many more solutions through

$$\sum_n D_n f_n(x,t),$$

where D_1, D_2, D_3, ... represent constants. Because any D_n can be any possible complex number, this is a big set of solutions; indeed it might be a big enough set to be the most general solution. Once we have the most general solution, we will need to find the values of D_n that correspond to the particular initial condition $\psi(x,0)$.

Casting about for special solutions: separation of variables. So, how do we find even one solution of the PDE? Let's try a solution $f(x,t)$ that is a product of a function $X(x)$ of position alone and a function $T(t)$ of time alone, that is, try a solution of the form

$$f(x,t) = X(x)T(t).$$

Plugging this guess into the PDE, we find

$$X(x)\frac{dT(t)}{dt} = -\frac{i}{\hbar}\left[-\frac{\hbar^2}{2m}\frac{d^2X(x)}{dx^2}T(t) + V(x)X(x)T(t)\right],$$

where the partial derivatives have become ordinary derivatives because they now act upon functions of a single variable. Divide both sides by $X(x)T(t)$ to find

$$\frac{1}{T(t)}\frac{dT(t)}{dt} = -\frac{i}{\hbar}\left[-\frac{\hbar^2}{2m}\frac{1}{X(x)}\frac{d^2X(x)}{dx^2} + V(x)\right].$$

In this equation, there is a function of time alone on the left-hand side, and a function of position alone on the right-hand side. But time and position are independent variables. It seems as if the left-hand side will vary with time, even while the position is held constant so the right-hand side stays constant! Similarly the other way around. There is only one way a function of t alone can always be equal to a function of x alone, and that's if both sides are equal to the same constant.

We don't yet know what that constant is, or how many such constants there might be. To allow for the possibility that there might be many such constants, we call the constant value of the quantity in square brackets by the name E_n. (This name suggests, correctly, that this constant must have the dimensions of energy.) We conclude that

$$\frac{1}{T(t)}\frac{dT(t)}{dt} = -\frac{i}{\hbar}E_n \tag{4.16}$$

$$-\frac{\hbar^2}{2m}\frac{1}{X(x)}\frac{d^2X(x)}{dx^2} + V(x) = E_n.$$

We started with one partial differential equation in two variables but (for solutions of the form $f(x,t) = X(x)T(t)$) ended with two ordinary differential equations. And we know a lot about how to solve ordinary differential equations! This technique for finding special solutions of the PDE is called "separation of variables".

Solving the first ODE. When faced with solving two equations, I always solve the easy one first. That way, if the result is zero, I won't have to bother solving the second equation.

So first we try to solve

$$\begin{aligned} \frac{1}{T(t)}\frac{dT(t)}{dt} &= -\frac{i}{\hbar}E_n \\ \frac{dT(t)}{T(t)} &= -\frac{i}{\hbar}E_n\,dt \\ \int \frac{dT}{T} &= -\frac{i}{\hbar}E_n \int dt \\ \ln T &= -\frac{i}{\hbar}E_n(t + \text{ constant }) \\ T_n(t) &= T_0 e^{-(i/\hbar)E_n t} \end{aligned}$$

Well, that went well. I don't know about you, but it was easier than *I* expected. In the last step, I changed the name of $T(t)$ to $T_n(t)$ to reflect the fact that we get different solutions for different values of E_n.

Solving the second ODE. We move on to

$$-\frac{\hbar^2}{2m}\frac{1}{X(x)}\frac{d^2X(x)}{dx^2} + V(x) = E_n.$$

Remembering the form of the infinite square well potential, and the boundary conditions $\psi(0,t) = 0$ plus $\psi(L,t) = 0$, the problem to solve is

$$-\frac{\hbar^2}{2m}\frac{d^2X(x)}{dx^2} = E_n X(x) \quad \text{with } X(0) = 0; X(L) = 0. \tag{4.17}$$

Perhaps you regard this sort of ordinary differential equation as unfair. After all, you don't yet know the permissible values of E_n. I'm not just asking you to solve an ODE with given coefficients, I'm asking you find find out what the coefficients are! Fair or not, we plunge ahead.

You are used to solving differential equations of this form. If I wrote

$$M\frac{d^2f(t)}{dt^2} = -kf(t),$$

you'd respond: "Of course, this is the ODE for a classical mass on a spring! The solution is

$$f(t) = C\cos(\omega t) + D\sin(\omega t) \quad \text{where } \omega = \sqrt{k/M}.\text{"}$$

Well, then, the solution for $X(x)$ has to be

$$X_n(x) = C_n\cos(\omega x) + D_n\sin(\omega x) \quad \text{where } \omega = \sqrt{2mE_n/\hbar^2},$$

where again I have taken to calling $X(x)$ by the name $X_n(x)$ to reflect the fact there there are different solutions for different values of E_n. Writing this out neatly,

$$X_n(x) = C_n\cos((\sqrt{2mE_n}/\hbar)x) + D_n\sin((\sqrt{2mE_n}/\hbar)x). \tag{4.18}$$

When you solved the classical problem of a mass on a spring, you had to supplement the ODE solution with the initial values $f(0) = x_0$, $f'(0) = v_0$, to find the constants C and D. This is called an "initial value problem". For the problem of a particle in a box, we don't have an initial value problem; instead we are given $X_n(0) = 0$ and $X_n(L) = 0$, which is called a "boundary value problem".

Plugging $x = 0$ into equation (4.18) will be easier than plugging in $x = L$, so I'll do that first. The result gives

$$X_n(0) = C_n \cos(0) + D_n \sin(0) = C_n,$$

so the boundary value $X_n(0) = 0$ means that $C_n = 0$ — for all values of n! Thus

$$X_n(x) = D_n \sin((\sqrt{2mE_n}/\hbar)x). \tag{4.19}$$

Now plug $x = L$ into equation (4.19), giving

$$X_n(L) = D_n \sin((\sqrt{2mE_n}/\hbar)L),$$

so the boundary value $X_n(L) = 0$ means that

$$\frac{\sqrt{2mE_n}}{\hbar}L = n\pi \qquad \text{where } n = 0, \pm 1, \pm 2, \pm 3, \ldots$$

and it follows that

$$X_n(x) = D_n \sin((n\pi/L)x).$$

If you think about it for a minute, you'll realize that $n = 0$ gives rise to $X_0(x) = 0$. True, this is a solution to the differential equation, but it's not an interesting one. Similarly, the solution for $n = -3$ is just the negative of the solution for $n = +3$, so we get the same effect by changing the sign of D_3. We don't have to worry about negative or zero values for n.

In short, the solutions for the boundary value problem are

$$X_n(x) = D_n \sin(n\pi x/L) \qquad \text{where } n = 1, 2, 3, \ldots$$

and with

$$E_n = n^2 \frac{\pi^2 \hbar^2}{2mL^2}.$$

We have accomplished the "unfair": we have not only solved the differential equation, we have also determined the permissible values of E_n.

Pulling things together. We now know that a solution to the Schrödinger time evolution equation for the infinite square well of width L is

$$\psi(x,t) = \sum_{n=1}^{\infty} D_n e^{-(i/\hbar)E_n t} \sin(n\pi x/L),$$

where

$$E_n = n^2 \frac{\pi^2 \hbar^2}{2mL^2}.$$

This is a lot of solutions — there are an infinite number of adjustable parameters D_n, after all! — but the question is whether it is the *most general* solution. In fact it *is* the most general solution, although that's not obvious. The branch of mathematics devoted to such questions, called Sturm-Liouville[5] theory, is both powerful and beautiful, but this is not the place to explore it.

Fitting the initial conditions. Remember that our problem is not simply to solve a PDE, it is to find how a given initial wavefunction

$$\psi(x,0) = \psi_0(x)$$

changes with time. To do this, we fit our solution to the given initial conditions.

To carry out this fitting, we must find D_n such that

$$\psi(x,0) = \sum_{n=1}^{\infty} D_n \sin(n\pi x/L) = \psi_0(x).$$

The problem seems hopeless at first glance, because there are an infinite number of unknowns D_n, yet only one equation! But there's a valuable trick, worth remembering, that renders it straightforward.

The trick relies on the fact that, for n, m integers,

$$\int_0^L \sin(n\pi x/L)\sin(m\pi x/L)\,dx = \begin{cases} L/2 & \text{for } n = m \\ 0 & \text{for } n \neq m \end{cases}. \tag{4.20}$$

You can work this integral out for yourself, using either

$$\sin A \sin B = \tfrac{1}{2}[\cos(A-B) - \cos(A+B)]$$

[5]Charles-François Sturm (1803–1855), French mathematician, also helped make the first experimental determination of the speed of sound in water. Joseph Liouville (1809–1882), another French mathematician, made contributions in complex analysis, number theory, differential geometry, and classical mechanics. He was also a public servant elected to the French Constituent Assembly of 1848, which established the Second Republic.

or else

$$\sin\theta = \frac{e^{+i\theta} - e^{-i\theta}}{2i},$$

whichever you like better. (Or you can look at problem 4.5, "Informal integration", on page 138, for an informal but easily remembered treatment.)

To employ this fact, start with

$$\sum_{n=1}^{\infty} D_n \sin(n\pi x/L) = \psi_0(x),$$

multiply both sides by $\sin(m\pi x/L)$, and integrate from 0 to L:

$$\sum_{n=1}^{\infty} D_n \int_0^L \sin(n\pi x/L)\sin(m\pi x/L)\,dx = \int_0^L \psi_0(x)\sin(m\pi x/L)\,dx.$$

This looks even worse, until you realize that all but one of the terms on the left vanish! Once you do make that realization, you find

$$D_m(L/2) = \int_0^L \psi_0(x)\sin(m\pi x/L)\,dx$$

and you have a formula for D_m.

Pulling all things together. For a particle of mass m ambivating in an infinite square well of width L, how does the quantal wave function change ("evolve") with time? If the initial wavefunction is $\psi_0(x)$, then the wavefunction at time t is

$$\psi(x,t) = \sum_{n=1}^{\infty} D_n e^{-(i/\hbar)E_n t}\sin(n\pi x/L), \tag{4.21}$$

where

$$E_n = n^2\frac{\pi^2\hbar^2}{2mL^2} \tag{4.22}$$

and

$$D_n = \frac{2}{L}\int_0^L \psi_0(x)\sin(n\pi x/L)\,dx. \tag{4.23}$$

Problem

4.5 **Informal integration** (recommended problem)
The integral (4.20) undergirds the Fourier sine series technique, and it's useful to remember. Here's how I do it. If $n \neq m$ the integrand is sometimes positive, sometimes negative over its range from 0 to L, so it's plausible that the two signs cancel out and result in a zero integral. If $n = m$ the integrand is always positive, so it must not be zero. But what is it?

a. Below is a graph of $\sin^2(3\pi x/L)$.

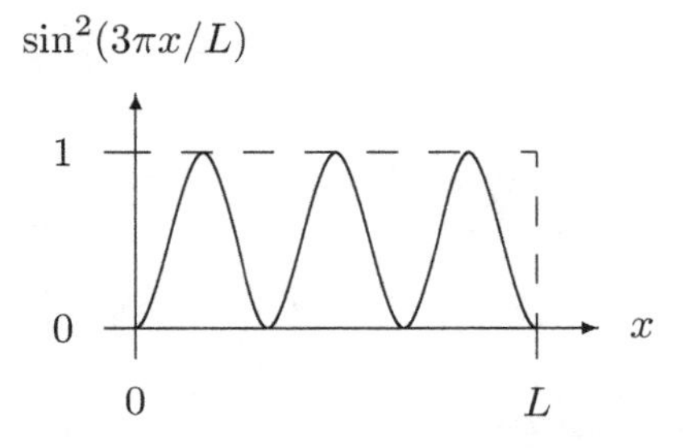

What is the area within the dashed box? Does it look like the area above the curve within the box is the same as the area below the curve within the box? What can you conclude about the value of the integral

$$\int_0^L \sin^2(3\pi x/L)\, dx\ ?$$

b. Make the above argument rigorous using the relationship

$$\sin^2\theta + \cos^2\theta = 1.$$

Your result should be that the integral

$$\int_0^L \sin^2(n\pi x/L)\, dx$$

has the same value whenever n is a non-zero integer.

4.6 What did we learn by solving the Schrödinger time evolution equation for the infinite square well?

In one sense, we learned that the time evolution of a particle of mass m in an infinite square well of width L, with initial wavefunction $\psi_0(x)$, is given by equation (4.21). But we should delve more deeply than simply saying "There's the answer, now let's go to sleep."

Quantal revivals. Does this jumble of symbols tell us anything about nature? Does it have any peculiar properties? Here's one. Suppose there were a time T_{rev} such that

$$e^{-(i/\hbar)E_n T_{\text{rev}}} = 1 \qquad \text{for } n = 1, 2, 3, \ldots. \tag{4.24}$$

What would the wavefunction $\psi(x,t)$ look like at time $t = T_{\text{rev}}$? It would be exactly equal to the initial wavefunction $\psi_0(x)$! If there is such a time, it's called the "revival time".

But it's not clear that such a revival time exists. After all, equation (4.24) lists an infinite number of conditions to be satisfied for revival to occur. Let's investigate. Because $e^{-i\,2\pi\,\text{integer}} = 1$ for any integer, the revival conditions (4.24) are equivalent to

$$(1/\hbar)E_n T_{\text{rev}} = 2\pi(\text{an integer}) \qquad \text{for } n = 1, 2, 3, \ldots.$$

Combined with the separation constant values (4.22), these conditions are

$$n^2 \frac{\pi\hbar}{4mL^2} T_{\text{rev}} = (\text{an integer}) \qquad \text{for } n = 1, 2, 3, \ldots.$$

And, looked at this way, it's clear that yes, there is a time T_{rev} that satisfies this infinite number of conditions. The smallest such time is

$$T_{\text{rev}} = \frac{4mL^2}{\pi\hbar}. \tag{4.25}$$

Cute and unexpected! This behavior is packed into equations (4.21) and (4.22), but no one would have uncovered this from a glance.

Moving across a node. Think about the wavefunction

$$D \sin(3\pi x/L).$$

This wavefunction and corresponding probability density are graphed below the infinite square well potential energy function.

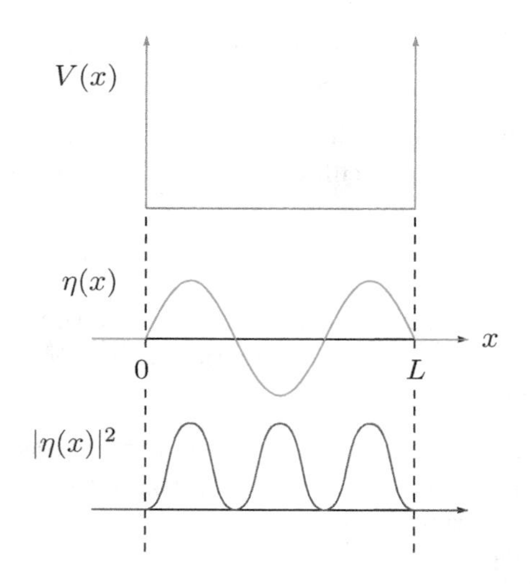

This particular wavefunction has two interior zeros, also called nodes. A common question is "There is zero probability of finding the particle at the node, so how can it move from one side of the node to the other?" People who ask this question suffer from the misconception that the particle is an infinitely small, infinitely hard version of a classical marble, which hence has a definite position. They think that the definite position of this infinitely small marble is changing rapidly, or changing erratically, or changing unpredictably, or changing subject to the slings and arrows of outrageous fortune. In truth, the quantal particle in this state doesn't have a definite position: it doesn't have a position at all! The quantal particle in the state above doesn't, *can't*, change its position from one side of the node to the other, because the particle *doesn't have a position.*

Investigating the solution technique. But I want to do more than investigate the properties of the solution, I want to investigate the characteristics of the solution technique. In his book *Mathematics in Action,* O. Graham Sutton writes that "A technique succeeds in mathematical physics, not by a clever trick, or a happy accident, but because it expresses some aspect of a physical truth." What aspect of physical truth is exposed through the techniques we developed to solve this time evolution problem?

First, let's review the problem we solved, then the techniques we used. The problem was solving the partial differential equation

$$\frac{\partial \psi(x,t)}{\partial t} = -\frac{i}{\hbar}\left[-\frac{\hbar^2}{2m}\frac{\partial^2 \psi(x,t)}{\partial x^2} + V(x)\psi(x,t)\right],$$

subject to the initial condition

$$\psi(x,0) = \psi_0(x).$$

The three techniques used were:

(1) Finding many particular solutions of the PDE that happen to factorize: $f(x,t) = X(x)T(t)$ ("separation of variables").
(2) Summing all of these particular solutions to find a more general (and, as it turns out, the most general) PDE solution: $\sum D_n X_n(x)T_n(t)$ ("superposition").
(3) Finding the coefficients D_n that match up to initial value $\psi_0(x)$ ("Fourier[6] sine series").

Fourier sine series. Let's look at the last step first. The technique of Fourier sine series is generally powerful. Any function $f(x)$ with $f(0) = 0$ and $f(L) = 0$ can be expressed as

$$f(x) = \sum_{n=1}^{\infty} f_n \sin(n\pi x/L) \qquad \text{where} \qquad f_n = \frac{2}{L}\int_0^L f(x)\sin(n\pi x/L)\,dx.$$

This seems paradoxical: complete information about the function is obtained through knowing $f(x)$ at every real number $0 \le x \le L$. Alternatively, complete information about the function is obtained through knowing the coefficients f_n for every positive integer n. But there are more real numbers between 0 and L than there are positive integers! I have no resolution for this paradox — I'll just remark that in knowing the function through its Fourier coefficients f_n, it seems that we're getting something for nothing.

[6] Joseph Fourier (1768–1830) was French, so his name is pronounced "Four - e - a" with a silent "r". He arrived at the series which today bears his name through studies of heat flow. He was the first to propose the phenomenon that we today call "the greenhouse effect". (So much for the climate-change denialist claim that the greenhouse effect is a modern day liberal/Chinese hoax.)

Well, there are lots of times when we want to get something for nothing! Fourier sine series are useful in data compression. For example, suppose you want to record a sound that starts with silence at time 0, proceeds through several notes, then ends with silence at time L. You could do this by keeping track of the air pressure $f(t)$ at every instant from 0 to L, or you could do it by keeping track of the corresponding Fourier coefficients f_n. In either case an infinite amount of data are required, so some will have to be thrown out to let it fit within a finite computer. It is more efficient to store this information in the form of f_n than in the form of $f(t)$: for a given amount of storage space, the f_n provide a more accurate reproduction of the sound than the $f(t)$. There are many schemes for the details of exactly when the Fourier series should be truncated: one such scheme is called "MP3".

Or, for pictures rather than sounds: A black-and-white photograph is a two-dimensional intensity function $f(x, y)$. You could store the image on a computer by breaking space (x, y) into a grid ("pixels") and storing a value for the intensity at each grid point (the so-called bitmap or BMP format) or you could store the information through Fourier coefficients $f_{n,m}$ (the so-called JPEG format). For a given level of image quality, the JPEG file is considerably smaller than the BMP file.

Stationary states. Okay, this is fun and profitable, but it tells us about how clever humans are; it doesn't tell us anything about nature. I'm going to probe in another direction: We see that, as far as time evolution is concerned, functions like $\sin(n\pi x/L)$ play a special role. What if the initial wavefunction $\psi_0(x)$ happens to have this form? We investigate $n = 3$. Once you see how things work in this case, you can readily generalize to any positive integer n.

So the initial wavefunction is

$$\psi_0(x) = A \sin(3\pi x/L).$$

We need the constant A so that the initial wavefunction will (1) have dimensions and (2) be normalized. For all wavefunctions, the probability of being *somewhere* is 1, that is

$$\int_{-\infty}^{+\infty} |\psi(x)|^2 \, dx = 1.$$

This requirement is called "normalization". Applying the general normalization requirement to this initial wavefunction for our particle in a box results in

$$\int_0^L A^2 \sin^2(3\pi x/L)\, dx = 1,$$

whence (remembering the sine-squared integral 4.20)

$$A^2(L/2) = 1 \quad \text{so} \quad A = \sqrt{2/L}.$$

Notice that then

$$\psi_0(x) = \sqrt{\frac{2}{L}} \sin(3\pi x/L)$$

has the proper dimensions.

Well, for this initial wavefunction, what are the values of

$$D_n = \frac{2}{L} \int_0^L \psi_0(x) \sin(n\pi x/L)\, dx \ ?$$

They are

$$\begin{aligned} D_n &= \frac{2}{L}\sqrt{\frac{2}{L}} \int_0^L \sin(3\pi x/L) \sin(n\pi x/L)\, dx \\ &= \frac{2}{L}\sqrt{\frac{2}{L}} \times \begin{cases} L/2 \text{ for } n = 3 \\ 0 \quad\ \text{for } n \neq 3 \end{cases} \\ &= \sqrt{\frac{2}{L}} \times \begin{cases} 1 \text{ for } n = 3 \\ 0 \text{ for } n \neq 3 \end{cases}, \end{aligned}$$

so

$$\psi(x,t) = \sqrt{\frac{2}{L}}\, e^{-(i/\hbar)E_3 t} \sin(3\pi x/L). \tag{4.26}$$

That's it! For this particular initial wavefunction, the system remains always in that same wavefunction, except multiplied by an time-dependent phase factor of $e^{-(i/\hbar)E_3 t}$. This uniform phase factor has *no effect whatsoever* on the probability density! Such states are called "stationary states".

Generic states. Contrast the time evolution of stationary states with the time evolution of generic states. For example, suppose the initial wavefunction were

$$\psi_0(x) = \frac{4}{5}\sqrt{\frac{2}{L}} \sin(3\pi x/L) + \frac{3}{5}\sqrt{\frac{2}{L}} \sin(7\pi x/L).$$

How does this state change with time? You should check two things: First, the wavefunction $\psi_0(x)$ given here is normalized. Second, it evolves in time to

$$\psi(x,t) = \frac{4}{5}\sqrt{\frac{2}{L}}\, e^{-(i/\hbar)E_3 t} \sin(3\pi x/L) + \frac{3}{5}\sqrt{\frac{2}{L}}\, e^{-(i/\hbar)E_7 t} \sin(7\pi x/L). \tag{4.27}$$

Although it takes a little effort to see exactly *how* the probability density changes with time, it's clear from a glance that it *does* change with time. This is *not* a stationary state.

Let's go back to the Einstein relation

$$E = \hbar\omega.$$

Neither Einstein nor de Broglie was ever clear about what it was that was "oscillating" with frequency ω, but now we have a better idea. In stationary state (4.26), the amplitude at every point oscillates with frequency $E_3/\hbar$. Using the Einstein relation, we say this state has energy E_3.

In contrast, the amplitude in generic state (4.27) has no single oscillation: there's a combination of frequency $E_3/\hbar$ and frequency $E_7/\hbar$. This state *doesn't have* an energy, in the same way that a silver atom with $\mu_x = +\mu_B$ *doesn't* have a value of μ_z, in the same way that an atom in state $|z+\rangle$ passing through a horizontal interferometer *doesn't have* a position, in the same way that love *doesn't have* a color. Instead, this state has amplitude $\frac{4}{5}$ to have energy E_3 and amplitude $\frac{3}{5}$ to have energy E_7.

We have uncovered the "aspect of physical truth" expressed by the separation constant E_n.

Energy eigenstates. How did the remarkable stationary states come about? Remember how they arose mathematically: we looked for solutions to

$$-\frac{\hbar^2}{2m}\frac{d^2X_n(x)}{dx^2} + V(x)X_n(x) = E_nX_n(x),$$

and the solutions we found (for the infinite square well) were those functions

$$X_n(x) = \sin(n\pi x/L)$$

that we later used as building blocks to build up *any* wavefunction. These now seem important enough that they warrant their own name. Because each is associated with a particularly energy E_n we call them "energy states". Because wavefunctions are usually represented by Greek letters we give them the name $\eta_n(x)$ where the Greek letter η (eta) suggests "energy" through alliteration. We write

$$-\frac{\hbar^2}{2m}\frac{d^2\eta_n(x)}{dx^2} + V(x)\eta_n(x) = E_n\eta_n(x), \tag{4.28}$$

and recognize this as one of those "unfair" problems where you must find not only the ODE solution $\eta_n(x)$, but you must also find the value of E_n. Such problems are given a half-German, half-English name: eigenproblems. We say that the "energy eigenfunction" $\eta_n(x)$ represents a "stationary state", or an "energy state", or an "energy eigenstate", and that E_n is an "energy eigenvalue". (The German word *eigen* derives from the same root as the English word "own", as in "my own state". It means "characteristic of" or "peculiar to" or "belonging to". The eigenstate $\eta_3(x)$ is the state "belonging to" energy E_3.)

The normalized energy eigenstate $\eta_3(x)$ is

$$\eta_3(x) = \sqrt{\frac{2}{L}} \sin(3\pi x/L).$$

We saw at equation (4.26) that this energy eigenstate evolves in time to

$$\psi(x,t) = e^{-(i/\hbar)E_3 t}\eta_3(x). \tag{4.29}$$

This state "belongs to" the energy E_3. In contrast, the state

$$\psi(x,t) = \tfrac{4}{5}e^{-(i/\hbar)E_3 t}\eta_3(x) + \tfrac{3}{5}e^{-(i/\hbar)E_7 t}\eta_7(x) \tag{4.30}$$

does *not* "belong to" any particular energy, because it involves both E_3 and E_7. Instead, this state has amplitude $\frac{4}{5}$ to have energy E_3 and amplitude $\frac{3}{5}$ to have energy E_7. We say that this state is a "superposition" of the energy states $\eta_3(x)$ and $\eta_7(x)$.

A particle trapped in a one-dimensional infinite square well cannot have any old energy: the only energies possible are the energy eigenvalues E_1, E_2, E_3, ... given in equation (4.22).

From the very first page of the very first chapter of this book we have been talking about quantization. But always before it has been a supplement added to the theory to make the results come out right: Max Planck added energy quantization to the otherwise-classical theory of blackbody radiation (equation 1.9); Niels Bohr supplemented the theory of a point-like classical electron orbiting a point-like classical proton with the requirement that the circular orbit contain a quantized (integer) number of de Broglie wavelengths (equation 1.26). Here, for the first time, quantization comes out of the theory, rather than being shoehorned into the beginning of the theory. Here, for the first time, the theory predicts that the only possible energies are those listed in equation (4.22). We have reached a milestone in our development of quantum mechanics.

Because the only possible energies are the energy eigenvalues E_1, E_2, E_3, ..., some people get the misimpression that the only possible *states* are the energy eigenstates $\eta_1(x)$, $\eta_2(x)$, $\eta_3(x)$, ... That's false. The state (4.30), for example, is a superposition of two energy states with different energies.

Analogy. A silver atom in magnetic moment state $|z+\rangle$ enters a vertical interferometer. It passes through the upper path. While traversing the interferometer, this atom has a position.

A different silver atom in magnetic moment state $|x-\rangle$ enters that same vertical interferometer. It ambivates through both paths. In more detail (see equation 3.19), it has amplitude $\langle z+|x-\rangle = -\frac{1}{\sqrt{2}}$ to take the upper path and amplitude $\langle z-|x-\rangle = \frac{1}{\sqrt{2}}$ to take the lower path, but it doesn't take a path. While traversing the interferometer, this atom has no position in the same way that love has no color.

A particle trapped in an infinite square well has state $\eta_6(x)$. This particle has energy E_6.

A different particle trapped in that same infinite square well has state

$$\frac{1}{\sqrt{2}}\eta_3(x) - \frac{1}{\sqrt{2}}\eta_4(x).$$

This particle does not have an energy. In more detail, it has amplitude $\frac{1}{\sqrt{2}}$ to have energy E_3 and amplitude $-\frac{1}{\sqrt{2}}$ to have energy E_4, but it doesn't have an energy in the same way that love doesn't have a color.

Summary. Our journey into quantum mechanics started with the experimental fact of quantized energies of blackbody radiation or of an atom. This inspired a search for quantized values of μ_z, which in turn prompted discovery of the new phenomena of interference and entanglement. Interference experiments suggested the mathematical tool of amplitude, and generalizing amplitude from magnetic moment to position prompted the mathematical tool of wavefunction. We asked the obvious question of how wavefunction changed with time, and answering that question brought us back to energy quantization with deeper insight. As T.S. Eliot wrote,

We shall not cease from exploration
And the end of all our exploring
Will be to arrive where we started
And know the place for the first time.

Problems

4.6 **Revival**

In an infinite square well, any wavefunction comes back to itself after the revival time given in equation (4.25) has passed. What happens after one-half of this time has passed?

4.7 **Normalized for all time**

Show that the wavefunction (4.27) is normalized for all values of the time t.

4.8 **Zero-point energy**

The lowest possible energy for a classical infinite square well is zero. The lowest possible energy for a quantal infinite square well is E_1 as given in equation (4.22). This difference is called the "zero-point energy" or the "vacuum energy".

Your grandparents are intelligent and thoughtful but have little background in science. They hold a bank trust fund for your eventual benefit. They have learned that on 27 May 2008, U.S. Patent 7379286 for "Quantum Vacuum Energy Extraction" was issued to the Jovion Corporation and they know that, if zero-point energy could be harnessed, it would produce enormous societal and financial gains. Your grandparents are thinking of withdrawing the trust fund money from the bank and investing it in the Jovion Corporation, but they want your advice before making the investment. What do you tell them? (Remember their intelligence: they want not just advice but concise, cogent reasoning behind that advice.)

4.9 **Fourier sine series for tent** (recommended problem)

Suppose the initial wavefunction is a pure real tent:

$$\psi_0(x) = \begin{cases} 0 & x < 0 \\ Ax & 0 \le x \le L/2 \\ A(L-x) & L/2 \le x \le L \\ 0 & L < x \end{cases}.$$

a. Sketch this initial wavefunction.

b. Show that, to insure normalization, we must use

$$A = \frac{2}{L}\sqrt{\frac{3}{L}}.$$

c. Verify that $\psi_0(x)$ has the proper dimensions.

d. The Fourier expansion coefficients are

$$\begin{aligned} D_n &= \frac{2}{L}\int_0^L \psi_0(x)\sin(n\pi x/L)\,dx \\ &= \frac{2A}{L}\left[\int_0^{L/2} x\sin(n\pi x/L)\,dx + \int_{L/2}^{L}(L-x)\sin(n\pi x/L)\,dx\right]. \end{aligned}$$

Before rushing in to evaluate this integral, pause to think! Without evaluating the integrals, show that when n is even the second integral within square brackets is the negative of the first integral, whereas when n is odd these two integrals are equal. Consequently, for this particular $\psi_0(x)$, when n is even $D_n = 0$, while when n is odd

$$D_n = \frac{4}{L}\int_0^{L/2} \psi_0(x)\sin(n\pi x/L)\,dx.$$

e. Now go ahead and evaluate D_n for n odd. (I used the substitution $\theta = n\pi x/L$, but there are other ways to do it.)

f. Write out the Fourier sine series representation for $\psi_0(x)$. Check that it has the proper dimensions, that it satisfies $\psi_0(0) = \psi_0(L) = 0$, and that it satisfies $\psi_0(L/2) = AL/2$. For this last check, use the result

$$1 + \frac{1}{3^2} + \frac{1}{5^2} + \frac{1}{7^2} + \cdots = \frac{\pi^2}{8}.$$

4.10 Find the flaw: Fourier sine series for ramp[7]

After working the above problem four students — Aldo, Beth, Celine, and Denzel — decide to find the Fourier sine series representation of the ramp wavefunction

$$\psi_0(x) = \begin{cases} 0 & x < 0 \\ Ax & 0 \le x < L \\ 0 & L \le x \end{cases}.$$

They split up to work independently, and when they get together afterwards they find that they have produced four different answers!

[7]Background concerning "find the flaw" type problems is provided in sample problem 1.2.1 on page 26.

Aldo: $\displaystyle -\frac{2}{\pi}\sqrt{\frac{3}{L^3}}\sum_{n=1}^{\infty}\frac{(-1)^n}{n}\sin(n\pi x/L)$

Beth: $\displaystyle -\frac{2}{\pi}\sqrt{\frac{3}{L}}\sum_{n=1}^{\infty}\frac{(-1)^n}{n}\cos(n\pi x/L)$

Celine: $\displaystyle -\frac{2}{\pi}\sqrt{\frac{3}{L}}\sum_{n=1,3,5,\cdots}\frac{(-1)^n}{n}\sin(n\pi x/L)$

Denzel: $\displaystyle -\frac{2}{\pi}\sqrt{\frac{3}{L}}\sum_{n=1}^{\infty}\frac{(-1)^n}{n}\sin(n\pi x/L)$

Provide simple reasons showing that three of these candidates must be wrong.

4.7 Other potentials

Is quantization peculiar to the infinite square well? No. At this stage in your mathematical education you don't have the tools to prove it, but in fact the infinite square well is entirely generic.

For any one-dimensional potential energy function that has $V(x) \to \infty$ when $x \to \pm\infty$, there are energy eigenstates $\eta_n(x)$ with discrete energy eigenvalues E_n such that

$$-\frac{\hbar^2}{2m}\frac{d^2\eta_n(x)}{dx^2} + V(x)\eta_n(x) = E_n\eta_n(x), \qquad n = 1, 2, 3, \ldots \tag{4.31}$$

and with the property that

$$\int_{-\infty}^{+\infty} \eta_m^*(x)\eta_n(x)\,dx = \begin{cases} 1 \text{ for } n = m \\ 0 \text{ for } n \neq m \end{cases}. \tag{4.32}$$

A side note on vocabulary is that this property is called "orthonormality". A side note on notation is that the symbol on the right-hand side of equation (4.32) defines the "Kronecker[8] delta":

$$\delta_{n,m} \equiv \begin{cases} 1 \text{ for } n = m \\ 0 \text{ for } n \neq m \end{cases}. \tag{4.33}$$

[8]Leopold Kronecker (1823–1891), German mathematician. After earning his Ph.D. he spent a decade managing a farm, which made him financially comfortable enough that he could pursue mathematics research for the rest of his life as a private scholar without university position.

Any wavefunction can be expressed as a "superposition" or "linear combination" of energy states:

$$\psi_0(x) = \sum_{n=1}^{\infty} c_n \eta_n(x) \tag{4.34}$$

where

$$c_n = \int_{-\infty}^{+\infty} \eta_n^*(x)\psi_0(x)\, dx. \tag{4.35}$$

We say that the energy states "span the set of wavefunctions" or that they constitute a "basis", but don't let these fancy terms bamboozle you: they just mean that starting from the energy states, you can use linear combinations to build up any wavefunction. The basis states in quantum mechanics play the same role as building blocks in a child's construction set. Just as a child can build castles, roadways, or trees — anything she wants — out of building blocks, so you can build any wavefunction you want out of energy states.

Superposition is the mathematical reflection of the physical phenomenon of interference. An example of quantal interference is "the atom passing through an interferometer doesn't take either path; instead it has amplitude c_a to take path a and amplitude c_b to take path b". An example of superposition is "the particle with wavefunction $\psi_0(x)$ given in equation (4.34) doesn't have an energy; instead it has amplitude c_n to have energy E_n, with $n = 1, 2, 3, \ldots$". It requires no great sophistication to see that these are parallel statements.

In the infinite square well, each energy eigenstate has a different energy. This is *not* always true: it might happen that two energy eigenvalues, perhaps E_8 and E_9, are equal. In this situation any linear combination

$$\alpha\, \eta_8(x) + \beta\, \eta_9(x) \tag{4.36}$$

is again an energy eigenstate with energy $E_8 = E_9$. (Although, to insure normalization, we must select $|\alpha|^2 + |\beta|^2 = 1$.) The two eigenvalues are then said to be "degenerate". I don't know how such a disparaging term[9] came to be attached to such a charming result, but it has been. If two eigenstates

[9] According to George F. Simmons, "This terminology follows a time-honored tradition in mathematics, according to which situations that elude simple analysis are dismissed by such pejorative terms as 'improper', 'inadmissible', 'degenerate', 'irregular', and so on." [*Differential Equations with Applications and Historical Notes*, third edition (CRC Press, Boca Raton, Florida, 2017) page 220.]

have the same energy, that energy is called "two-fold degenerate". If three have the same energy, it is "three-fold degenerate". And so forth.

Finally, solving the energy eigenproblem opens the door to solving the time evolution problem, because the wavefunction $\psi_0(x)$ evolves in time to

$$\psi(x,t) = \sum_{n=1}^{\infty} c_n e^{-(i/\hbar)E_n t} \eta_n(x). \tag{4.37}$$

Because the energy eigenproblem (4.31) tells you the resulting quantized energy values, and because energy quantization is one of the easiest aspects of quantum mechanics to access experimentally, some people develop the mistaken impression that finding energy values is all there is to quantum mechanics. No. It's actually just like a classical mechanics problem: "You stand atop a cliff 97 meters tall and hurl a ball horizontally at 12 m/s. (a) How far from the base of the cliff does it land? (b) At what speed does it strike the ground?" Using energy techniques alone you can answer question (b). But to answer both questions you need to solve the time evolution problem. The same holds in quantum mechanics. Some questions can be answered knowing only the energy eigenvalues, but to answer *any* question you must solve the time evolution problem. Often the easiest way to do this is by first solving the energy eigenproblem (finding both eigenvalues and eigenfunctions) and then employing the time evolution equation (4.37). The energy eigenvalues are important — no doubt about that — but they're not the full story.

Problem

4.11 **Basis with degeneracy** (essential problem)

You know that if the two orthonormal vectors $\left\{\hat{\imath}, \hat{\jmath}\right\}$ constitute a basis for position vectors in two dimensions, then rotating the pair by angle θ produces two different orthonormal vectors

$$\begin{aligned} \hat{\imath}' &= \cos\theta\,\hat{\imath} + \sin\theta\,\hat{\jmath} \\ \hat{\jmath}' &= -\sin\theta\,\hat{\imath} + \cos\theta\,\hat{\jmath} \end{aligned} \tag{4.38}$$

that constitute a basis just as good as the original basis.

Show that if

$$\eta_n(x) \qquad n = 1, 2, 3, \ldots$$

constitutes an orthonormal basis of energy eigenstates (an "energy eigenbasis"), with a degeneracy so that $E_8 = E_9$, then if $\eta_8(x)$ and $\eta_9(x)$ are replaced by

$$\begin{aligned} \eta_8'(x) &= \quad \cos\theta\, \eta_8(x) + \sin\theta\, \eta_9(x) \\ \eta_9'(x) &= -\sin\theta\, \eta_8(x) + \cos\theta\, \eta_9(x) \end{aligned} \tag{4.39}$$

this new basis is just as good an orthonormal energy eigenbasis as the original.

4.8 Energy loss

I said earlier [at equations (4.26) and (4.29)] that any energy eigenstate $\eta_6(x)$ is a "stationary state": that if the system started off in state $\eta_6(x)$, it would remain in that state forever (with a time-dependent phase factor in front). This seems to contradict the experimental fact that most of the atoms we find lying about are in their ground states.[10] Why don't they just stay in state $\eta_6(x)$ for ever and ever?

Furthermore: If the system starts off in the state $c_3\eta_3(x)+c_7\eta_7(x)$, then for all time the probability of measuring energy E_3 is $|c_3|^2$, the probability of measuring energy E_7 is $|c_7|^2$, and the probability of measuring the ground state energy is zero. Again, how can this conclusion be consistent with the experimental observation that most atoms are in the ground state?

The answer is that if time evolution *were* given exactly by equation (4.12),

$$\frac{\partial \psi(x,t)}{\partial t} = -\frac{i}{\hbar}\left[-\frac{\hbar^2}{2m}\frac{\partial^2 \psi(x,t)}{\partial x^2} + V(x)\psi(x,t)\right], \tag{4.40}$$

then the atom *would* stay in that stationary state forever. But real atoms are subject to collisions and radiation meaning that the time-evolution equation above is *not* exactly correct. These phenomena, unaccounted for in the equation above, cause the atom to fall into its ground state.

Because collisions and radiation are small effects, an atom starting off in state $\eta_6(x)$ stays in that stationary state for a "long" time — but that means long relative to typical atomic times, such as the characteristic time 10^{-17} seconds generated at problem 1.15 on page 41. If you study more

[10]The energy eigenstate with lowest energy eigenvalue has a special name: the ground state.

quantum mechanics,[11] you will find that a typical atomic excited state lifetime is 10^{-9} seconds. So the excited state lifetime is very short by human standards, but very long by atomic standards. (To say "very long" is an understatement: it is 100 million times longer; by contrast the Earth has completed only 66 million orbits since the demise of the dinosaurs.)

4.9 Mean values

Recall the interference experiment. A single atom ambivates through the two paths of an interferometer.

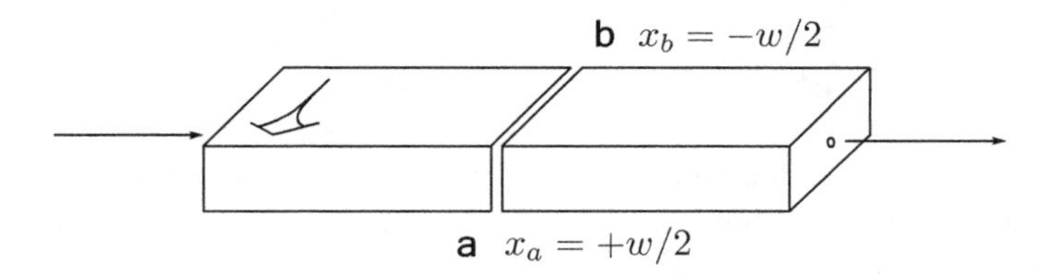

Say the interferometer has width w, so that path a has position $x_a = +w/2$ while path b has position $x_b = -w/2$.

You know the drill: the atom has amplitude c_a of taking path a, amplitude c_b of taking path b. If a lamp is turned on while an interference experiment is proceeding, the probability of the atom appearing in path a is $|c_a|^2$, the probability of the atom appearing in path b is $|c_b|^2$. In other words, if the atom's position is measured while the interference experiment is proceeding, the result would be $+w/2$ with probability $|c_a|^2$, and it would be $-w/2$ with probability $|c_b|^2$. Hence the mean position measured would be

$$+ (w/2)|c_a|^2 - (w/2)|c_b|^2. \qquad (4.41)$$

It seems weird to say "The atom doesn't have a position but its mean position is given by equation (4.41)" — sort of like saying "Unicorns don't exist but their mean height is 1.3 meters." Indeed, it would be more accurate to say "The atom doesn't have a position but if a light were turned on

[11] See for example David J. Griffiths and Darrell F. Schroeter, *Introduction to Quantum Mechanics*, third edition (Cambridge University Press, Cambridge, UK, 2018) section 11.3.2, "The Lifetime of an Excited State".

— or if position could be determined in some other way — then it would have a position, and the mean position found in that way would be given by equation (4.41)." This more accurate sentence is such a mouthful that it's rarely said: people say the first, inaccurate sentence as shorthand for the second, correct sentence.

A particle in a potential. A particle is in state

$$\psi(x,t) = \sum_{n=1}^{\infty} c_n e^{-(i/\hbar)E_n t}\eta_n(x). \tag{4.42}$$

Here c_n is the amplitude that the particle has energy E_n, so $|c_n|^2$ is the probability that, if the energy were measured, the result E_n would be found. The mean energy is thus clearly

$$\langle E\rangle = \sum_{n=1}^{\infty} |c_n|^2 E_n. \tag{4.43}$$

Once again, it seems weird to have a formula for the mean energy of a particle that doesn't have an energy. The meaning is that if the energy were measured, $\langle E\rangle$ is the mean of the energy that would be found.

New expression for mean energy. The above expression for mean energy is correct but difficult to use. Suppose a particle with wavefunction $\psi(x)$ is subject to a potential energy function $V(x)$. To find the mean energy you must: (1) Solve the energy eigenproblem to find the energy eigenvalues E_n and eigenfunctions $\eta_n(x)$. (2) Write the wavefunction $\psi(x)$ in the form $\psi(x) = \sum_n c_n\eta_n(x)$. (3) Now that you know the energies E_n and the amplitudes (expansion coefficients) c_n, execute the sum $\sum_n |c_n|^2 E_n$. Whew! Isn't there an easier way?

There is. The wavefunction (4.42) has time derivative

$$\frac{\partial\psi(x,t)}{\partial t} = \sum_{n=1}^{\infty} c_n \left(-\frac{i}{\hbar}E_n\right) e^{-(i/\hbar)E_n t}\eta_n(x) \tag{4.44}$$

and complex conjugate

$$\psi^*(x,t) = \sum_{m=1}^{\infty} c_m^* e^{+(i/\hbar)E_m t}\eta_m^*(x). \tag{4.45}$$

Thus

$$\begin{aligned}&\psi^*(x,t)\frac{\partial\psi(x,t)}{\partial t} \\ &\quad = \sum_{m=1}^{\infty}\sum_{n=1}^{\infty} c_m^* c_n \left(-\frac{i}{\hbar}E_n\right) e^{-(i/\hbar)(E_n-E_m)t}\eta_m^*(x)\eta_n(x)\end{aligned} \tag{4.46}$$

and

$$\int_{-\infty}^{+\infty} \psi^*(x,t)\frac{\partial \psi(x,t)}{\partial t}\,dx \tag{4.47}$$
$$= \sum_{m=1}^{\infty}\sum_{n=1}^{\infty} c_m^* c_n \left(-\frac{i}{\hbar}E_n\right) e^{-(i/\hbar)(E_n-E_m)t} \int_{-\infty}^{+\infty} \eta_m^*(x)\eta_n(x)\,dx.$$

But (see equation 4.32) the integral on the right is zero unless $m = n$, in which case it is 1. Thus

$$\int_{-\infty}^{+\infty} \psi^*(x,t)\frac{\partial \psi(x,t)}{\partial t}\,dx = \sum_{n=1}^{\infty} c_n^* c_n \left(-\frac{i}{\hbar}E_n\right) = -\frac{i}{\hbar}\langle E\rangle \tag{4.48}$$

or

$$-\frac{i}{\hbar}\langle E\rangle = \int_{-\infty}^{+\infty} \psi^*(x,t)\left(\frac{\partial \psi(x,t)}{\partial t}\right)dx. \tag{4.49}$$

Now, according to the Schrödinger equation

$$\frac{\partial \psi(x,t)}{\partial t} = -\frac{i}{\hbar}\left[-\frac{\hbar^2}{2m}\frac{\partial^2\psi(x,t)}{\partial x^2} + V(x)\psi(x,t)\right], \tag{4.50}$$

so

$$\langle E\rangle = \int_{-\infty}^{+\infty} \psi^*(x,t)\left[-\frac{\hbar^2}{2m}\frac{\partial^2\psi(x,t)}{\partial x^2} + V(x)\psi(x,t)\right]dx. \tag{4.51}$$

This expression for mean energy does not require solving the energy eigenproblem or expanding $\psi(x)$ in energy eigenstates.

This expression parallels the expressions already determined in problem 4.2 on page 125: For example the mean position is

$$\langle x\rangle = \int_{-\infty}^{+\infty} \psi^*(x,t)\,[x]\,\psi(x,t)\,dx, \tag{4.52}$$

the mean of position squared is

$$\langle x^2\rangle = \int_{-\infty}^{+\infty} \psi^*(x,t)\,\left[x^2\right]\psi(x,t)\,dx, \tag{4.53}$$

and indeed for any function of position $f(x)$, the mean is

$$\langle f(x)\rangle = \int_{-\infty}^{+\infty} \psi^*(x,t)\,[f(x)]\,\psi(x,t)\,dx. \tag{4.54}$$

Problems

4.12 Mean position vs. "expected position"

For the infinite square well energy eigenstate $\eta_2(x)$, what is the mean position? What is the probability density at that point? Is this mean position really the "expected position"?

4.13 Wavefunction vs. probability density (recommended problem)

The wavefunction $\psi(x)$ is not directly measurable, but can be inferred (up to an overall phase) through a number of position and interference experiments. The probability density $|\psi(x)|^2$ is measurable through a number of position experiments alone. These facts lead some to the misconception that the probability density tells the "whole story" of a quantal state. This problem demonstrates the falsehood of that misconception by presenting a series of wavefunctions, all with the same probability density, but each with a different mean energy. (And hence each with different behavior in the future.) The so-called Gaussian wavefunctions are

$$\psi(x) = Ae^{-x^2/2\sigma^2}e^{ikx},$$

where A is a normalization constant.

a. (Mathematical preliminary.) Use integration by parts to show that

$$\int_{-\infty}^{+\infty} e^{-t^2}\, dt = 2\int_{-\infty}^{+\infty} t^2 e^{-t^2}\, dt,$$

where t is any dimensionless variable.

b. When the particle is free, $V(x) = 0$, find the mean energy. (In this case the mean kinetic energy, since there is no potential energy.) If you use the above result, you will not need to evaluate any integral nor find the normalization constant.

4.14 Mean of function vs. function of mean (recommended problem)

Show that $\langle f(x)\rangle$ might not equal $f(\langle x\rangle)$ by using the function $f(x) = x^2$ and the so-called Gaussian wavefunction

$$\psi(x) = Ae^{-x^2/2\sigma^2}e^{ikx}.$$

a. What is the normalization constant A? Does your answer have the proper dimensions?
b. What is the mean position $\langle x\rangle$ for this wavefunction?
c. What is the mean of the function $\langle f(x)\rangle$?
d. What is the function of the mean $f(\langle x\rangle)$?

4.10 The classical limit of quantum mechanics

I told you way back on page 4 that when quantum mechanics is applied to big things, it gives the results of classical mechanics. It's hard to see how my claim could possibly be correct: the whole structure of quantum mechanics differs so dramatically from the structure of classical mechanics — the character of a "state", the focus on potential energy function rather than on force, the emphasis on energy eigenproblems instead of initial value problems, the fact that the quantal time evolution equation involves a first derivative with respect to time while the classical time evolution equation involves a second derivative with respect to time.

4.10.1 *How does mean position change with time?*

This nut is cracked by focusing, not on the full quantal state $\psi(x,t)$, but on the mean position

$$\langle x \rangle = \int_{-\infty}^{+\infty} \psi^*(x,t) x \psi(x,t)\, dx, \tag{4.55}$$

How does this mean position change with time?

The answer depends on the classical force function $F(x)$ — i.e., the classical force that would be exerted on a classical particle if it were at position x. (I'm not saying that the particle *is* at x, I'm not even saying that the particle has a position; I'm saying that's what the force *would be* if the particle *were* classical and at position x.)

The answer is that

$$\langle F(x) \rangle = m \frac{d^2 \langle x \rangle}{dt^2}, \tag{4.56}$$

a formula that certainly plucks our classical heartstrings! This result is called *Ehrenfest's theorem.*[12] We will prove this theorem later (in section 4.10.4 on page 162), but first discuss its significance.

Although the theorem is true in all cases, it is most useful when the spread in position Δx is in some sense small, so the wavefunction is relativity compact. Such wavefunctions are called "wavepackets". In this

[12]Paul Ehrenfest (1880–1933) contributed to relativity theory, quantum mechanics, and statistical mechanics, often by pointing out concrete difficulties in these theories. As a result, several telling arguments have names like "Ehrenfest's paradox" or "Ehrenfest's urn" or "the Ehrenfest dog-flea model".

situation we might hope for a useful approximation — the classical limit — by ignoring the quantal indeterminacy of position and focusing solely on mean position.

If the force function $F(x)$ varies slowly on the scale of Δx, then our hopes are confirmed: the spread in position is small, the spread in force is small, and to a good approximation the mean force $\langle F(x) \rangle$ is equal to the force at the mean position $F(\langle x \rangle)$.

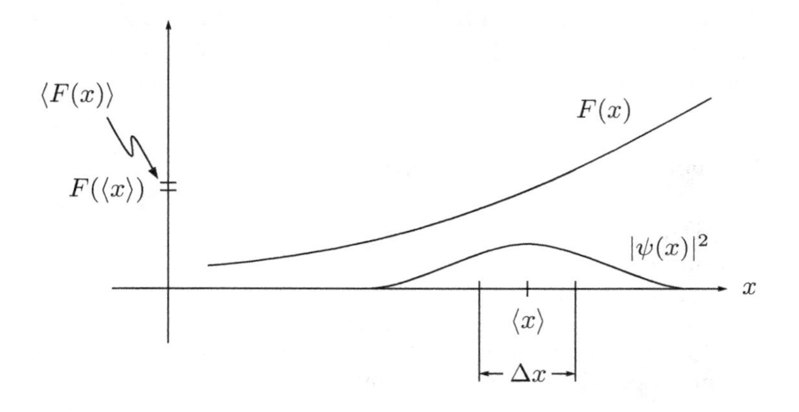

But if the force function varies rapidly on the scale of Δx, then our hopes are dashed: the spread in position is small, but the spread in force is not, and the classical approximation is not appropriate.

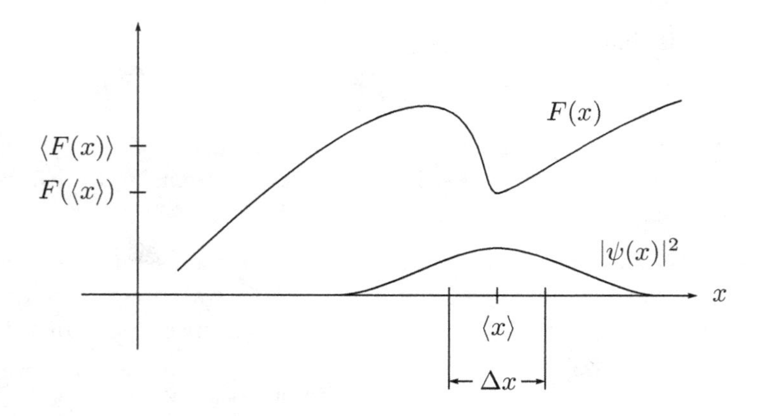

To head off a misconception, I emphasize that Ehrenfest's theorem is *not* that

$$F(\langle x \rangle) = m\frac{d^2\langle x \rangle}{dt^2}.$$

If this were true, then the mean position of a quantal particle would in all cases move exactly as a classical particle does. But (see problem 4.14, "Mean of function vs. function of mean", on page 156) it's *not* true.

4.10.2 *Is the classical approximation good enough?*

If the quantal position indeterminacy Δx is small compared to the experimental uncertainty of your position-locating experimental apparatus, for the entire duration of your experiment, then the classical approximation is usually appropriate. So the central question is: How big is the quantal Δx in my situation? This will of course vary from case to case and from time to time within a given case. But there's an important theorem that connects the indeterminacy of position Δx with the indeterminacy of momentum Δp: in *all* situations

$$\Delta x \Delta p \geq \tfrac{1}{2}\hbar. \tag{4.57}$$

This theorem is called the *Heisenberg indeterminacy principle.* Because this book has focused on position and not momentum, we cannot prove the theorem at this time: you'll have to read a more advanced book. But you should know about the result for two reasons: First, because it's important for determining whether the classical limit is appropriate in a given case. Second, because it was important in the historical development of quantum mechanics.

Quantum mechanics has a long and intricate (and continuing!) history, but one of the keystone events occurred in the spring of 1925. Werner Heisenberg,[13] a freshly minted Ph.D., had obtained a position as assistant

[13]German theoretical physicist (1901–1976) who nearly failed his Ph.D. oral exam due to his fumbling in experimental physics. He went on to discover quantum mechanics as we know it today. Although attacked by Nazis as a "white Jew", he became a principal scientist in the German nuclear program during World War II, where he focused on building nuclear reactors rather than nuclear bombs. After the war he worked to rebuild German science, and to extend quantum theory into relativistic and field theoretic domains. He enjoyed hiking, particularly in the Bavarian Alps, and playing the piano. After a three-month whirlwind romance, Heisenberg married Elisabeth Schumacher, sister of the *Small Is Beautiful* economist E.F. Schumacher, and they went on to parent seven children.

to Max Born at the University of Göttingen. There he realized that the key to formulating quantum mechanics was to develop a theory that fit the experiments described in chapter 1, and that also had the correct classical limit. He was searching for such a theory when he came down with a bad case of allergies to spring pollen from the "mass of blooming shrubs, rose gardens and flower beds"[14] of Göttingen. He decided to travel to Helgoland, a rocky island and fishing center in the North Sea, far from pollen sources, arriving there by ferry on 8 June 1925.

Once his health returned, Heisenberg reproduced his earlier work, cleaning up the mathematics and simplifying the formulation. He worried that the mathematical scheme he invented might prove to be inconsistent, and in particular that it might violate the principle of energy conservation. In Heisenberg's own words:[15]

> One evening I reached the point where I was ready to determine the individual terms in the energy table, or, as we put it today, in the energy matrix, by what would now be considered an extremely clumsy series of calculations. When the first terms seemed to accord with the energy principle, I became rather excited, and I began to make countless arithmetical errors. As a result, it was almost three o'clock in the morning before the final result of my computations lay before me. The energy principle had held for all the terms, and I could no longer doubt the mathematical consistency and coherence of the kind of quantum mechanics to which my calculations pointed. At first, I was deeply alarmed. I had the feeling that, through the surface of atomic phenomena, I was looking at a strangely beautiful interior, and felt almost giddy at the thought that I now had to probe this wealth of mathematical structures nature had so generously spread out before me. I was far too excited to sleep, and so, as a new day dawned, I made for the southern tip of the island, where I had been longing to climb a rock jutting out into the sea. I now did so without too much trouble, and waited for the sun to rise.

Because the correct classical limit was essential in producing this theory, it was easy to fall into the misconception that an electron really did behave classically, with a single position, but that this single position is disturbed

[14] Werner Heisenberg, *Physics and Beyond* (Harper and Row, New York, 1971) page 37.
[15] *Physics and Beyond*, page 61.

by the measuring apparatus used to determine position. Indeed, Heisenberg wrote as much:[16]

> observation of the position will alter the momentum by an unknown and undeterminable amount.

But Neils Bohr repeatedly objected to this "disturbance" interpretation. For example, at a 1938 conference in Warsaw,[17] he

> warned specifically against phrases, often found in the physical literature, such as "disturbing of phenomena by observation."

Today, interference and entanglement experiments make clear that Bohr was right and that "measurement disturbs the system" is not a tenable position.[18] In an interferometer, there is *no local way* that a photon at path a can physically disturb an atom taking path b. For an entangled pair of atoms, there is *no local way* that an analyzer measuring the magnetic moment of the left atom can physically disturb the right atom. It is no defect in our measuring apparatus that it cannot determine what does not exist.

And this brings us to one last terminology note. What we have called the "Heisenberg indeterminacy principle" is called by some the "Heisenberg uncertainty principle".[19] The second name is less accurate because it gives the mistaken impression that an electron *really does* have a position and we are just uncertain as to what that position is. It also gives the mistaken impression that an electron *really does* have a momentum and we are just uncertain as to what that momentum is.

---

[16]Werner Heisenberg, *The Physical Principles of the Quantum Theory*, translated by Carl Eckart and F.C. Hoyt (University of Chicago Press, Chicago, 1930) page 20.

[17]Niels Bohr, "Discussion with Einstein on epistemological problems in atomic physics," in *Albert Einstein, Philosopher–Scientist*, edited by Paul A. Schilpp (Library of Living Philosophers, Evanston, Illinois, 1949) page 237.

[18]To be completely precise, "measurement disturbs the system locally" is not a tenable position. The "de Broglie–Bohm pilot wave" formulation of quantum mechanics can be interpreted as saying that "measurement disturbs the system", but the measurement at one point in space is felt instantly at points arbitrarily far away. When this formulation is applied to a two-particle system, a "pilot wave" situated in six-dimensional configuration space somehow physically guides the two particles situated in ordinary three-dimensional space.

[19]Heisenberg himself, writing in German, called it the "Genauigkeit Beziehung" — accuracy relationship. See "Über den anschaulichen Inhalt der quantentheoretischen Kinematik und Mechanik" *Zeitschrift für Physik* **43** (March 1927) 172–198.

4.10.3 *Sample Problem*

For the "Underground Guide to Quantum Mechanics" (described on page 20), you decide to write a passionate persuasive paragraph or two concerning the misconception that "measurement disturbs the system". What do you write?

Solution: For those of us who know and love classical mechanics, there's a band-aid, the idea that "measurement disturbs the system". This idea is that fundamentally classical mechanics actually holds, but that quantum mechanics is a mask layered over top of, and obscuring the view of, the classical mechanics because our measuring devices disturb the underlying classical system. That's not possible. It is no defect of our measuring instruments that they cannot determine what does not exist, just as it is no defect of a colorimeter that it cannot determine the color of love.

This idea that "measurement disturbs the system" is a psychological trick to comfort us, and at the same time to keep us from exploring, fully and openly, the strange world of quantum mechanics. I urge you, I implore you, to discard this security blanket, to go forth and discover the new world as it really is rather than cling to the familiar classical world. Like Miranda in Shakespeare's *Tempest*, take delight in this "brave new world, that has such people in't".

Unlike most band-aids, this band-aid does not protect or cover up. Instead it exposes a lack of imagination.

4.10.4 *Proof of Ehrenfest's theorem*

I'm not going to kid you: this derivation is long, difficult, and, frankly, unenlightening. It is necessary to show the coherence of the entire quantal scheme we've been building, and if you follow it critically you will learn some tricks of the trade, but if you decide to skip this section I won't be offended.

Because quantum mechanics emphasizes potential energy $V(x)$, and classical mechanics emphasizes force $F(x)$, let's recall how they're related. The definition of potential energy (in one dimension) is

$$V(x) - V(x_0) = -\int_{x_0}^{x} F(x')\,dx', \tag{4.58}$$

where $F(x)$ is the classical force function. Taking the derivative of both sides with respect to x (and using the fundamental theorem of calculus on the right-hand side) gives

$$\frac{\partial V(x)}{\partial x} = -F(x). \tag{4.59}$$

Remember that the classical time evolution equation is

$$F(x(t)) = m\frac{d^2x(t)}{dt^2}$$

which is second-order with respect to time and which, of course, contains no reference to $\hbar$.

We ask how the mean position moves in quantum mechanics:

$$\langle x\rangle = \int_{-\infty}^{+\infty} x\psi^*(x,t)\psi(x,t)\,dx, \tag{4.60}$$

so

$$\frac{d\langle x\rangle}{dt} = \int_{-\infty}^{+\infty} x\frac{\partial\psi^*(x,t)}{\partial t}\psi(x,t)\,dx + \int_{-\infty}^{+\infty} x\psi^*(x,t)\frac{\partial\psi(x,t)}{\partial t}\,dx. \tag{4.61}$$

But the Schrödinger time evolution equation tells us how wavefunction $\psi(x,t)$ changes with time:

$$\frac{\partial\psi(x,t)}{\partial t} = -\frac{i}{\hbar}\left[-\frac{\hbar^2}{2m}\frac{\partial^2\psi(x,t)}{\partial x^2} + V(x)\psi(x,t)\right] \tag{4.62}$$

and

$$\frac{\partial\psi^*(x,t)}{\partial t} = +\frac{i}{\hbar}\left[-\frac{\hbar^2}{2m}\frac{\partial^2\psi^*(x,t)}{\partial x^2} + V(x)\psi^*(x,t)\right]. \tag{4.63}$$

(From here on I'm going to write $\psi(x,t)$ as ψ and $V(x)$ as V.) Thus

$$\begin{aligned}
\frac{d\langle x\rangle}{dt} &= \frac{i}{\hbar}\left\{\int_{-\infty}^{+\infty} x\left[-\frac{\hbar^2}{2m}\frac{\partial^2\psi^*}{\partial x^2} + V\psi^*\right]\psi\,dx\right. \\
&\qquad \left. - \int_{-\infty}^{+\infty} x\psi^*\left[-\frac{\hbar^2}{2m}\frac{\partial^2\psi}{\partial x^2} + V\psi\right]dx\right\} \\
&= \frac{i}{\hbar}\left\{-\frac{\hbar^2}{2m}\left[\int_{-\infty}^{+\infty} x\frac{\partial^2\psi^*}{\partial x^2}\psi\,dx - \int_{-\infty}^{+\infty} x\psi^*\frac{\partial^2\psi}{\partial x^2}\,dx\right]\right. \\
&\qquad \left. + \int_{-\infty}^{+\infty} xV\psi^*\psi\,dx - \int_{-\infty}^{+\infty} x\psi^*V\psi\,dx\right\} \\
&= -i\frac{\hbar}{2m}\left[\int_{-\infty}^{+\infty} x\frac{\partial^2\psi^*}{\partial x^2}\psi\,dx - \int_{-\infty}^{+\infty} x\psi^*\frac{\partial^2\psi}{\partial x^2}\,dx\right].
\end{aligned} \tag{4.64}$$

It seems odd that the two terms involving potential energy cancel, so no explicit dependence on $V(x)$ appears in this result, but we'll just push on.

Can we say anything about integrals such as the second integral in square brackets above? Surprisingly, the answer is yes. If we define

$$f(x) = x\psi^* \qquad \text{and} \qquad g(x) = \frac{\partial \psi}{\partial x} \tag{4.65}$$

then

$$\int_{-\infty}^{+\infty} x\psi^* \frac{\partial^2 \psi}{\partial x^2}\, dx = \int_{-\infty}^{+\infty} f(x)g'(x)\, dx \tag{4.66}$$

which suggests integration by parts:

$$\int_{-\infty}^{+\infty} f(x)g'(x)\, dx = \Big[f(x)g(x)\Big]_{-\infty}^{+\infty} - \int_{-\infty}^{+\infty} f'(x)g(x)\, dx. \tag{4.67}$$

Now remember that the wavefunction is normalized, so it has to fall to zero at both infinity and negative infinity. Typically the slope $\partial\psi/\partial x$ also falls to zero at both infinity and negative infinity, and does so very rapidly — much more rapidly than linearly. (There are exceptions to these typical behaviors, such as scattering wavefunctions, and in these atypical cases this argument has to be rethought.) The upshot is that in typical situations

$$\Big[f(x)g(x)\Big]_{-\infty}^{+\infty} = 0 \tag{4.68}$$

so

$$\int_{-\infty}^{+\infty} x\psi^* \frac{\partial^2 \psi}{\partial x^2}\, dx = -\int_{-\infty}^{+\infty} \frac{\partial(x\psi^*)}{\partial x}\frac{\partial \psi}{\partial x}\, dx. \tag{4.69}$$

We'll use this trick several times. . . I'll just call it the "integration-by-parts trick".

Applying this trick to both integrals of equation (4.64) gives

$$\begin{aligned}
\frac{d\langle x\rangle}{dt} &= -i\frac{\hbar}{2m}\left[-\int_{-\infty}^{+\infty} \frac{\partial(x\psi)}{\partial x}\frac{\partial \psi^*}{\partial x}\, dx + \int_{-\infty}^{+\infty} \frac{\partial(x\psi^*)}{\partial x}\frac{\partial \psi}{\partial x}\, dx\right] \\
&= -i\frac{\hbar}{2m}\left[-\int_{-\infty}^{+\infty} x\frac{\partial \psi}{\partial x}\frac{\partial \psi^*}{\partial x}\, dx - \int_{-\infty}^{+\infty} \psi\frac{\partial \psi^*}{\partial x}\, dx\right. \\
&\qquad\qquad \left. + \int_{-\infty}^{+\infty} x\frac{\partial \psi^*}{\partial x}\frac{\partial \psi}{\partial x}\, dx + \int_{-\infty}^{+\infty} \psi^*\frac{\partial \psi}{\partial x}\, dx\right] \\
&= -i\frac{\hbar}{2m}\left[-\int_{-\infty}^{+\infty} \psi\frac{\partial \psi^*}{\partial x}\, dx + \int_{-\infty}^{+\infty} \psi^*\frac{\partial \psi}{\partial x}\, dx\right] \\
&= \frac{\hbar}{m}\Im m\left\{\int_{-\infty}^{+\infty} \psi^*\frac{\partial \psi}{\partial x}\, dx\right\}.
\end{aligned} \tag{4.70}$$

Notice that $d\langle x\rangle/dt$ is *pure real*, as it must be. And notice that the dimensions are the same on both sides. (This isn't proof that we've made no algebra errors, but if our expression for $d\langle x\rangle/dt$ had been complex, or if it had been dimensionally incorrect, then that *would have been* proof that we *had* made algebra errors.)

All this is fine and good, but it takes us only part way to our goal. This is clearly *not* a classical equation... it contains $\hbar$ right there! Since the classical $F = ma$ involves the second derivative of position with respect to time, we take one more time derivative of $\langle x\rangle$, finding

$$\frac{d^2\langle x\rangle}{dt^2} = \frac{\hbar}{m}\Im m\left\{\int_{-\infty}^{+\infty} \frac{\partial\psi^*}{\partial t}\frac{\partial\psi}{\partial x}\,dx + \int_{-\infty}^{+\infty} \psi^*\frac{\partial}{\partial x}\frac{\partial\psi}{\partial t}\,dx\right\}. \tag{4.71}$$

The second-order derivative on the right looks particularly grotesque, so we use the integration-by-parts trick to get rid of it:

$$\begin{aligned}\frac{d^2\langle x\rangle}{dt^2} &= \frac{\hbar}{m}\Im m\left\{\int_{-\infty}^{+\infty} \frac{\partial\psi^*}{\partial t}\frac{\partial\psi}{\partial x}\,dx - \int_{-\infty}^{+\infty} \frac{\partial\psi^*}{\partial x}\frac{\partial\psi}{\partial t}\,dx\right\}\\ &= -\frac{2\hbar}{m}\Im m\left\{\int_{-\infty}^{+\infty} \frac{\partial\psi^*}{\partial x}\frac{\partial\psi}{\partial t}\,dx\right\}.\end{aligned} \tag{4.72}$$

Now use the Schrödinger equation:

$$\begin{aligned}\frac{d^2\langle x\rangle}{dt^2} &= -\frac{2\hbar}{m}\Im m\left\{\int_{-\infty}^{+\infty} \frac{\partial\psi^*}{\partial x}\left\{-\frac{i}{\hbar}\left[-\frac{\hbar^2}{2m}\frac{\partial^2\psi}{\partial x^2} + V\psi\right]\right\}dx\right\}\\ &= \frac{2}{m}\Re e\left\{\int_{-\infty}^{+\infty} \frac{\partial\psi^*}{\partial x}\left[-\frac{\hbar^2}{2m}\frac{\partial^2\psi}{\partial x^2} + V\psi\right]dx\right\}.\end{aligned} \tag{4.73}$$

Look at that... two of the $\hbar$s have canceled out! We're not home yet because there's still an $\hbar$ within the square brackets, but we're certainly making progress. We have that

$$\frac{d^2\langle x\rangle}{dt^2} = \frac{2}{m}\Re e\left\{-\frac{\hbar^2}{2m}\int_{-\infty}^{+\infty} \frac{\partial\psi^*}{\partial x}\frac{\partial^2\psi}{\partial x^2}\,dx + \int_{-\infty}^{+\infty} \frac{\partial\psi^*}{\partial x}V\psi\,dx\right\}, \tag{4.74}$$

but let's apply the integration-by-parts trick to the first integral:

$$\int_{-\infty}^{+\infty} \frac{\partial\psi^*}{\partial x}\frac{\partial^2\psi}{\partial x^2}\,dx = -\int_{-\infty}^{+\infty} \frac{\partial^2\psi^*}{\partial x^2}\frac{\partial\psi}{\partial x}\,dx. \tag{4.75}$$

Think about this for a minute: if the integral on the left is z, this equation says that $z = -z^*$, whence z is pure imaginary or $\Re e\{z\} = 0$. Thus

$$\frac{d^2\langle x\rangle}{dt^2} = \frac{2}{m}\Re e\left\{\int_{-\infty}^{+\infty} \frac{\partial\psi^*}{\partial x}V\psi\,dx\right\}, \tag{4.76}$$

an expression devoid of $\hbar$s! Apply the integration-by-parts trick to this integral:

$$\begin{aligned}
\int_{-\infty}^{+\infty} \frac{\partial \psi^*}{\partial x} V\psi\, dx &= -\int_{-\infty}^{+\infty} \psi^* \frac{\partial (V\psi)}{\partial x}\, dx \\
\int_{-\infty}^{+\infty} \frac{\partial \psi^*}{\partial x} V\psi\, dx &= -\int_{-\infty}^{+\infty} \psi^* V \frac{\partial \psi}{\partial x}\, dx - \int_{-\infty}^{+\infty} \psi^* \frac{\partial V}{\partial x} \psi\, dx \\
\int_{-\infty}^{+\infty} \frac{\partial \psi^*}{\partial x} V\psi\, dx + \int_{-\infty}^{+\infty} \psi^* V \frac{\partial \psi}{\partial x}\, dx &= -\int_{-\infty}^{+\infty} \psi^* \frac{\partial V}{\partial x} \psi\, dx \\
2\Re e \left\{ \int_{-\infty}^{+\infty} \frac{\partial \psi^*}{\partial x} V\psi\, dx \right\} &= -\int_{-\infty}^{+\infty} \psi^* \frac{\partial V}{\partial x} \psi\, dx. \qquad (4.77)
\end{aligned}$$

Plugging this result back into equation (4.76) gives

$$\frac{d^2\langle x\rangle}{dt^2} = -\frac{1}{m}\int_{-\infty}^{+\infty} \psi^* \frac{\partial V}{\partial x} \psi\, dx. \qquad (4.78)$$

But the force function is $F(x) = -\partial V/\partial x$, so

$$\frac{d^2\langle x\rangle}{dt^2} = \frac{1}{m}\int_{-\infty}^{+\infty} \psi^*(x,t) F(x) \psi(x,t)\, dx = \frac{1}{m}\langle F(x)\rangle. \qquad (4.79)$$

There it is —

$$\langle F(x)\rangle = m\frac{d^2\langle x\rangle}{dt^2}. \qquad (4.80)$$

We have proven Ehrenfest's theorem.

4.11 Transitions induced by light

This section is more intricate than other sections of this book, and it takes many steps to reach its conclusion. Furthermore, it is not needed as background for any following section, so you might want to skip over it. But the steps are valuable and the conclusion itself is one of the most fascinating and useful relations in all of physics.

The problem. An electron in the ground state of a symmetric potential well is exposed to light. What is the probability that it transitions through light absorption to a particular excited state?

Expectation. I expect that the light would induce transitions from the ground state to the excited state. If a collection of atoms is exposed to light for two seconds, there will be twice as many transitions as there were when those atoms were exposed to light for one second.

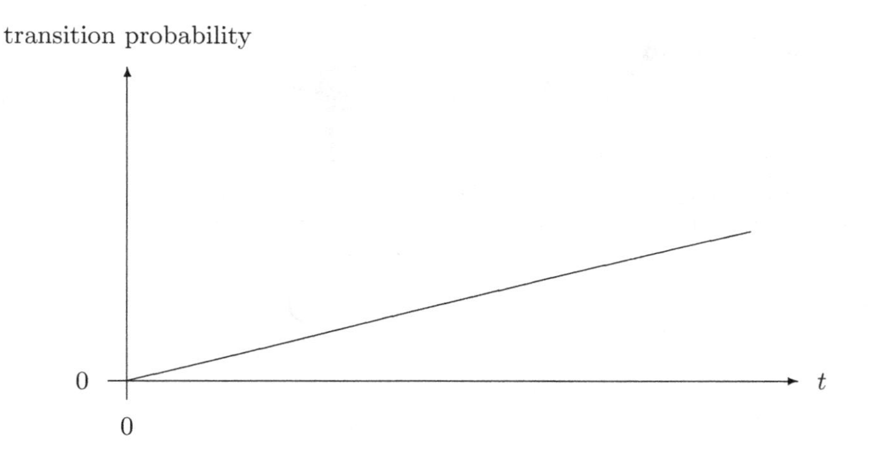

Setup. Call the ground state wavefunction $\eta_g(x)$ with energy E_g and the excited state wavefunction $\eta_e(x)$ with energy E_e. In light of the Einstein relation (1.21) we define the frequency characteristic of this transition

$$\omega_0 = (E_e - E_g)/\hbar. \tag{4.81}$$

For most symmetric potential wells the mean position for any energy eigenstate vanishes,

$$\langle x \rangle_g = 0 \quad \text{and} \quad \langle x \rangle_e = 0, \tag{4.82}$$

and we will assume this. (If this assumption is wrong, the derivation needs to be rethought.)

When no light shines, the Schrödinger time evolution equation is

$$\begin{aligned} \frac{\partial \psi(x,t)}{\partial t} &= -\frac{i}{\hbar}\left[-\frac{\hbar^2}{2m}\frac{\partial^2 \psi(x,t)}{\partial x^2} + V_{\text{well}}(x)\psi(x,t)\right] \\ &\equiv -\frac{i}{\hbar}\left[\mathcal{H}_{\text{well}}\psi(x,t)\right]. \end{aligned} \tag{4.83}$$

In this section we abbreviate the term in square brackets as "$\mathcal{H}_{\text{well}}\psi(x,t)$". For example, when $\psi(x,t) = \eta_g(x)$ we have

$$\mathcal{H}_{\text{well}}\eta_g(x) = E_g\eta_g(x), \tag{4.84}$$

and when $\psi(x,t) = \eta_e(x)$ we have

$$\mathcal{H}_{\text{well}}\eta_e(x) = E_e\eta_e(x). \tag{4.85}$$

You know how this time evolution behaves: the initial wavefunction

$$c_g\eta_g(x) + c_e\eta_e(x) \tag{4.86}$$

evolves in time to

$$c_g e^{-(i/\hbar)E_g t}\eta_g(x) + c_e e^{-(i/\hbar)E_e t}\eta_e(x). \tag{4.87}$$

When light *does* shine, the electron is subject not only to the well's potential energy function, but also to the potential energy function due to the light. If the electric field at the center of the well is $E_0\cos(\omega t)$, then that additional potential energy function is

$$eE_0\cos(\omega t)\,x, \tag{4.88}$$

where the charge on an electron is $-e$. In this circumstance the wavefunction no longer evolves like (4.87), but instead like

$$\psi(t) = c_g(t)e^{-(i/\hbar)E_g t}\eta_g(x) + c_e(t)e^{-(i/\hbar)E_e t}\eta_e(x). \tag{4.89}$$

Our job is to find the probability of starting in the ground state and ending in the excited state. That is, assuming $c_g(0) = 1$ and $c_e(0) = 0$, we need to find $c_e(t)$. The transition probability is then $|c_e(t)|^2$.

Time evolution when the light shines. When light shines, the potential energy function changes from that of the well alone, $V_{\text{well}}(x)$, to $V_{\text{well}}(x) + eE_0\cos(\omega t)\,x$. Thus the Schrödinger time evolution equation changes from equation (4.83) to

$$\frac{d}{dt}\psi(t) = -\frac{i}{\hbar}\left[\mathcal{H}_{\text{well}}\psi(t) + eE_0\cos(\omega t)\,x\psi(t)\right]. \tag{4.90}$$

First, look at the left-hand side:

$$\begin{aligned}\frac{d}{dt}\psi(t) &= \frac{d}{dt}\left[c_g(t)e^{-(i/\hbar)E_g t}\eta_g(x) + c_e(t)e^{-(i/\hbar)E_e t}\eta_e(x)\right] \qquad (4.91)\\ &= \left[\dot{c}_g(t)e^{-(i/\hbar)E_g t}\eta_g(x) + \dot{c}_e(t)e^{-(i/\hbar)E_e t}\eta_e(x)\right]\\ &\quad - \frac{i}{\hbar}\left[E_g c_g(t)e^{-(i/\hbar)E_g t}\eta_g(x) + E_e c_e(t)e^{-(i/\hbar)E_e t}\eta_e(x)\right].\end{aligned}$$

Meanwhile, on the right-hand side

$$\begin{aligned}\mathcal{H}_{\text{well}}\psi(t) &= c_g(t)e^{-(i/\hbar)E_g t}\mathcal{H}_{\text{well}}\eta_g(x) + c_e(t)e^{-(i/\hbar)E_e t}\mathcal{H}_{\text{well}}\eta_e(x)\\ &= c_g(t)e^{-(i/\hbar)E_g t}E_g\eta_g(x) + c_e(t)e^{-(i/\hbar)E_e t}E_e\eta_e(x). \qquad (4.92)\end{aligned}$$

Putting these three equations together gives

$$\begin{aligned} &\dot{c}_g(t)e^{-(i/\hbar)E_g t}\eta_g(x) + \dot{c}_e(t)e^{-(i/\hbar)E_e t}\eta_e(x) \\ &= -\frac{i}{\hbar}eE_0\cos(\omega t)\,x\left[c_g(t)e^{-(i/\hbar)E_g t}\eta_g(x) + c_e(t)e^{-(i/\hbar)E_e t}\eta_e(x)\right]. \end{aligned} \tag{4.93}$$

Multiply the above equation by $\eta_e^*(x)$ and integrate over all values of x. Because of orthonormality

$$\int_{-\infty}^{+\infty}\eta_e^*(x)\eta_g(x)\,dx = 0 \quad \text{and} \quad \int_{-\infty}^{+\infty}\eta_e^*(x)\eta_e(x)\,dx = 1, \tag{4.94}$$

while because $\langle x\rangle_e = 0$

$$\int_{-\infty}^{+\infty}\eta_e^*(x)\,x\,\eta_e(x)\,dx = 0. \tag{4.95}$$

Hence we find

$$\dot{c}_e(t)e^{-(i/\hbar)E_e t} = -\frac{i}{\hbar}eE_0\cos(\omega t)c_g(t)e^{-(i/\hbar)E_g t}\int_{-\infty}^{+\infty}\eta_e^*(x)\,x\,\eta_g(x)\,dx. \tag{4.96}$$

The integral on the right is just a (complex) number — not a function of x, not a function of t — and we'll call that number $\langle e|x|g\rangle$. Recalling definition (4.81), we write

$$\dot{c}_e(t) = -\frac{i}{\hbar}eE_0\langle e|x|g\rangle\cos(\omega t)e^{+i\omega_0 t}c_g(t). \tag{4.97}$$

If we had multiplied (4.93) instead by $\eta_g^*(x)$ and integrated we would have found

$$\dot{c}_g(t) = -\frac{i}{\hbar}eE_0\langle g|x|e\rangle\cos(\omega t)e^{-i\omega_0 t}c_e(t). \tag{4.98}$$

So far, we have made assumptions but no approximations.

Approximate solution of the time evolution equations. Our job is to find the transition probability $|c_e(t)|^2$ with the initial conditions $c_g(0) = 1$ and $c_e(0) = 0$. This coupled problem is difficult. However, we can make progress under conditions where $c_g(t)$ and $c_e(t)$ change *slowly.* In this case replace $c_g(t)$ on the right-hand side of (4.97) with its initial value, namely 1. (This approximation is called "first order perturbation theory for the time evolution problem".) This gives

$$\dot{c}_e(t) = -\frac{i}{\hbar}eE_0\langle e|x|g\rangle e^{i\omega_0 t}\cos(\omega t) \tag{4.99}$$

which integrates with respect to time to

$$c_e(t) = -\frac{i}{\hbar} e E_0 \langle e|x|g\rangle \int_0^t e^{i\omega_0 t'} \cos(\omega t')\, dt'. \tag{4.100}$$

Performing the integral gives

$$\begin{aligned}
\int_0^t e^{i\omega_0 t'} \cos(\omega t')\, dt' &= \tfrac{1}{2} \int_0^t e^{i\omega_0 t'} (e^{+i\omega t'} + e^{-i\omega t'})\, dt' \\
&= \frac{1}{2} \int_0^t \left[e^{i(\omega_0+\omega)t'} + e^{i(\omega_0-\omega)t'} \right] dt' \\
&= \frac{1}{2} \left[\frac{e^{i(\omega_0+\omega)t'}}{i(\omega_0+\omega)} + \frac{e^{i(\omega_0-\omega)t'}}{i(\omega_0-\omega)} \right]_0^t \\
&= \frac{1}{2i} \left[\frac{e^{i(\omega_0+\omega)t} - 1}{\omega_0+\omega} + \frac{e^{i(\omega_0-\omega)t} - 1}{\omega_0-\omega} \right].
\end{aligned}$$

This full expression is formidable, to put it mildly. But Enrico Fermi[20] noticed something about the magnitudes concerned. Look at

$$\frac{e^{i(\omega_0+\omega)t} - 1}{\omega_0+\omega}.$$

The numerator is a complex number with magnitude between 0 and 2. The denominator is a real number involving ω which, for light, is about $10^{15}\ \mathrm{s}^{-1}$. So this fraction will be numerically tiny. Similarly for the piece

$$\frac{e^{i(\omega_0-\omega)t} - 1}{\omega_0-\omega}$$

except that when $\omega \approx \omega_0$, the denominator is near zero so the fraction can be large indeed. In this way Fermi realized that the transition probability is almost always tiny. It is more than tiny only when $\omega \approx \omega_0$, and in that regime the approximation

$$\int_0^t e^{i\omega_0 t'} \cos(\omega t')\, dt' \approx \frac{1}{2i} \frac{e^{i(\omega_0-\omega)t} - 1}{\omega_0-\omega} \tag{4.101}$$

[20]Enrico Fermi (1901–1954) of Italy and the United States excelled in both experimental and theoretical physics. He directed the building of the first nuclear reactor and produced the first theory of the weak nuclear interaction. The Fermi surface in the physics of metals was named in his honor. He elucidated the statistics of what are now called fermions in 1926. He produced so many thoughtful conceptual and estimation problems that such problems are today called "Fermi problems". I never met him (he died before I was born) but I have met several of his students, and each of them speaks of him in that rare tone reserved for someone who is not just a great scientist and a great teacher and a great leader, but also a great human being.

is highly accurate.

Furthermore,

$$\frac{1}{2i}\frac{e^{i(\omega_0-\omega)t}-1}{\omega_0-\omega}=\frac{1}{2i}\frac{[e^{i(\omega_0-\omega)t/2}-e^{-i(\omega_0-\omega)t/2}]e^{i(\omega_0-\omega)t/2}}{\omega_0-\omega} \tag{4.102}$$

which seems like a step backwards, until you remember that $e^{i\theta}-e^{-i\theta}=2i\sin\theta$, so

$$\frac{1}{2i}\frac{e^{i(\omega_0-\omega)t}-1}{\omega_0-\omega}=\frac{\sin[(\omega_0-\omega)t/2]}{\omega_0-\omega}e^{i(\omega_0-\omega)t/2}. \tag{4.103}$$

So, at this excellent level of approximation,

$$c_e(t)=-\frac{i}{\hbar}eE_0\langle e|x|g\rangle\frac{\sin[(\omega_0-\omega)t/2]}{\omega_0-\omega}e^{i(\omega_0-\omega)t/2} \tag{4.104}$$

and the transition probability is

$$|c_e(t)|^2=\frac{e^2E_0^2|\langle e|x|g\rangle|^2}{\hbar^2}\frac{\sin^2[(\omega_0-\omega)t/2]}{(\omega_0-\omega)^2}. \tag{4.105}$$

Given all the assumptions and approximations we introduced to derive this result, you might think it's an obscure equation of limited applicability. You'd be wrong. It is used so often that Fermi called it the "golden rule".

Reflection. The transition probability result, graphed below as a function of time, shows oscillatory behavior called "Rabi[21] flopping". This is the beat at the heart of an atomic clock.

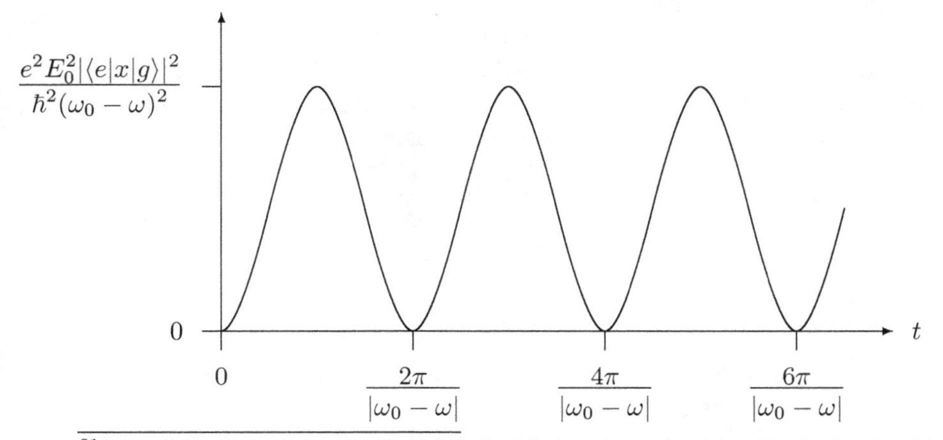

[21]Isidor Isaac Rabi (1898–1988), Polish-Jewish-American physicist. His fascinating life cannot be summarized in a few sentences: I recommend John Rigden's biography *Rabi: Scientist and Citizen* (Basic Books, New York, 1987).

I have made bad guesses in my life, but none worse than the difference between my expectation graphed on page 167 and the real behavior graphed above. It's as if, while hammering a nail into a board, the first few strikes drive the nail deeper and deeper into the board, but additional strikes make the nail come out of the board. And one strike (at time $2\pi/|\omega_0 - \omega|$) makes the nail pop out of the board altogether! Is there any way to account for this bizarre result other than shrugging that "It comes out of the math"?

There is. This is a form of interference[22] where the particle moves not from point to point through two possible slits, but from energy state to energy state with two possible intermediate states. The initial state is the ground state and the final state is the ground state. The two possible intermediates are the excited state and the ground state. There is an amplitude to go from ground state to ground state via the excited state, and an amplitude to go from ground state to ground state via the ground state. At time $\pi/|\omega_0 - \omega|$ those two amplitudes interfere destructively so there is a small probability of ending up in the ground state and hence a large probability of ending up in the excited state. At time $2\pi/|\omega_0 - \omega|$ those two amplitudes interfere constructively so there is a large probability of ending up in the ground state and hence a small probability of ending up in the excited state.

Problem

4.15 **Explore some more**

There's a lot more to say to flesh out the story told by equation (4.105), but I'll restrict myself to one question: The denominator vanishes when $\omega = \omega_0$, so you might think that the probability goes to infinity there. Bad idea. Show that the probability is instead

$$\frac{e^2 E_0^2 |\langle e|x|g\rangle|^2}{\hbar^2} \frac{t^2}{4}.$$

4.12 Position plus spin

In chapters 2 and 3 we investigated particles with magnetic moment, like the silver atom, doing our best to treat the quantum mechanics of mag-

[22]This point of view is expounded by R.P. Feynman and A.R. Hibbs in section 6-5 of *Quantum Mechanics and Path Integrals*, emended edition (Dover Publications, Mineola, NY, 2010).

netic moment while ignoring the quantum mechanics of position. Then in chapter 4 we investigated the quantum mechanics of position, ignoring the quantum mechanics of magnetic moment. For historical reasons, the magnetic moment is said to reflect the intrinsic "spin" of a particle. It's time to weld the two pieces of spin and position together.

This is achieved in a straightforward way. Think back to our discussion of bin amplitudes for a single particle in one dimension. We asked the question "What is the amplitude for the particle to be found in bin i?" But if the particle has a two-basis-state spin, like a silver atom, we have to ask the question "What is the amplitude for the particle to be found in bin i with spin up?" Or "What is the amplitude for the particle to be found in bin i with spin down?" (Alternatively we might ask for x-spin positive or x-spin negative, or $17°$-spin positive or negative, or for the projection on any other axis, but it's conventional to focus on the vertical axis.) We'll call the answer to the first question $\psi_{i,+}$, the answer to the second question $\psi_{i,-}$, and we'll write the two answers together as

$$\psi_i(+\tfrac{1}{2}) = \psi_{i,+} \quad \text{and} \quad \psi_i(-\tfrac{1}{2}) = \psi_{i,-}$$

so that the answer to both questions at once is $\psi_i(m_s)$, where $m_s = \pm\frac{1}{2}$. (The choice $m_s = \pm\frac{1}{2}$ instead of $m_s = \pm 1$ or even $m_s = \pm 3\pi$ is again for historical reasons.)

Now do the standard thing: divide by the square root of bin size and take the limit as bin size shrinks to zero. This quantity becomes an amplitude density (wavefunction)

$$\psi(x, m_s) \quad \text{where } -\infty < x < +\infty \text{ and } m_s = \pm\tfrac{1}{2}.$$

Sometimes people write this as two separate wavefunctions:

$$\psi_+(x) = \psi(x, +\tfrac{1}{2}) \quad \text{and} \quad \psi_-(x) = \psi(x, -\tfrac{1}{2}).$$

And sometimes they write it as a "spatial part" times a "spin part":

$$\psi_+(x) = \phi(x)\chi_+ \quad \text{and} \quad \psi_-(x) = \xi(x)\chi_-.$$

But don't let the notation fool you: all of these expressions represent the same thing.

It might happen that the two spatial parts are equal, $\phi(x) = \xi(x)$, in which case we can say

$$\psi(x, m_s) = \phi(x)\chi(m_s) \quad \text{where } \chi(+\tfrac{1}{2}) = \chi_+ \text{ and } \chi(-\tfrac{1}{2}) = \chi_-.$$

But not all wavefunctions factorize in this way.

If the atom were nitrogen, with four possible spin projections (see page 45), then we would have to ask "What is the amplitude for the nitrogen atom to be found in bin i with vertical spin projection $+\frac{3}{2}$?" or $+\frac{1}{2}$ or $-\frac{1}{2}$ or $-\frac{3}{2}$. (Alternatively we might ask "What is the amplitude for the nitrogen atom to be found in bin i with a projection on the 67° axis of $-\frac{3}{2}$?") After asking these questions and taking the appropriate limits, the relevant wavefunction will be

$$\psi(x, m_s) \quad \text{where } -\infty < x < +\infty \text{ and } m_s = +\tfrac{3}{2}, +\tfrac{1}{2}, -\tfrac{1}{2}, -\tfrac{3}{2}.$$

In the same way for an atom of sulfur, with five possible spin projections, the relevant wavefunction will be

$$\psi(x, m_s) \quad \text{where } -\infty < x < +\infty \text{ and } m_s = +2, +1, 0, -1, -2.$$

If the atom moves in three dimensions, the wavefunction will take the form

$$\psi(x, y, z, m_s) \equiv \psi(\mathsf{x}), \tag{4.106}$$

where the single sans serif symbol x stands for the four variables x, y, z, m_s. [Because the variables x, y, and z are continuous, while the variable m_s is discrete, one sometimes sees the dependence on m_s written as a subscript rather than as an argument: $\psi_{m_s}(x, y, z)$. This is a bad habit: m_s is a variable not a label, and it should not be notated as a second-class variable just because it's discrete.]

The electron is a spin-$\frac{1}{2}$ particle. So are the positron and the muon, and all the quarks. Protons and neutrons are composite particles, made up of quarks, but the quarks combine in such a way that the proton and the neutron are spin-$\frac{1}{2}$ particles. The He^3 atom is also a composite particle, made up of protons, neutrons, and electrons, but in its ground state it has spin $\frac{1}{2}$. The same applies, as we've seen extensively in chapters 3 and 4, for the silver atom.

The Higgs boson is a spin-0 particle. So is the composite He^4 atom (in its ground state). The photon is a spin-1 particle, but it does not behave like a typical massive spin-1 particle because it is intrinsically relativistic.

Epilogue

The quantum mechanics of position is very strange, yes.
And it's very difficult, yes.
But it's also very wonderful.

Problems

4.16 **Spin-$\frac{1}{2}$ electron in a potential well, I** (essential problem)
All electrons are spin-$\frac{1}{2}$ particles. Using the χ_+, χ_- notation, write down the wavefunction for an electron ambivating in a potential well with energy eigenfunctions $\eta_n(x)$:

a. with amplitude $\frac{4}{5}$ of being in the spatial ground state ($n = 1$) with spin up and amplitude $\frac{3}{5}$ of being in the spatial $n = 3$ state with spin down;
b. with amplitude $\frac{4}{5}$ of being in the spatial ground state ($n = 1$) with spin up and amplitude $\frac{3}{5}$ of being in the spatial $n = 3$ state with negative spin projection on the x axis (see equation 3.19).

4.17 **Spin-$\frac{1}{2}$ electron in a potential well, II**
An electron ambivates in a potential well with with energy eigenvalues E_n and energy eigenfunctions $\eta_n(x)$. The electron's wavefunction is

$$\tfrac{3}{5}\eta_3(x)\chi_+ + \tfrac{4}{5}\eta_4(x)\chi_-.$$

a. What is the mean energy?
b. If the vertical spin projection is measured, what is the probability of finding +? (That is, of finding $m_s = +\frac{1}{2}$.)
c. The vertical spin projection is measured and found to be +. Now what is the mean energy?

4.18 **Questions** (recommended problem)
Update your list of quantum mechanics questions that you started at problem 1.17 on page 46. Write down new questions and, if you have uncovered answers to any of your old questions, write them down briefly.

⟦For example, one of my questions would be: "What are the detailed mechanisms for the energy loss outlined in section 4.8?"⟧

Chapter 5

Solving the Energy Eigenproblem

Energy eigenproblems are important: they determine the "allowed" energy eigenvalues, and, as chapter 1 made clear, such energy quantization is the most experimentally accessible facet of quantum mechanics. Also, the easiest way to solve the time evolution problem is to first solve the energy eigenproblem. This chapter focuses only on the spatial part of the wavefunction and ignores any spin part. For particles with spin, the two parts can be welded together using the techniques of section 4.12 on page 172.

In fact, Erwin Schrödinger discovered the energy eigenproblem first (in December 1925) and five months later discovered the time evolution equation, which he called "the true wave equation". Today, both equations carry the name "Schrödinger equation", which can result in confusion.

There are large numbers of analytic and numerical techniques for solving eigenproblems. Most of these are effective but merely technical: they find the answer, but don't give any insight into the character of the resulting energy eigenfunctions. For example, if you study more quantum mechanics you will find that for the simple harmonic oscillator, $V(x) = \frac{1}{2}kx^2$, the energy eigenfunctions are

$$\eta_n(x) = \left(\frac{\sqrt{km}/\hbar}{2^{2n}(n!)^2\pi} \right)^{1/4} e^{-(\sqrt{km}/2\hbar)x^2} H_n((\sqrt{km}/\hbar)^{1/2}x) \qquad (5.1)$$
$$\text{for } n = 0, 1, 2, 3, \ldots$$

where the Hermite polynomials are defined through

$$H_n(z) = (-1)^n e^{z^2} \frac{d^n e^{-z^2}}{dz^n}.$$

Yikes! This is true, but provides little insight.

This chapter presents two of the many solution techniques available. First we investigate an informal, rough-and-ready technique for sketching energy eigenfunctions that doesn't give rigorous solutions, but that does provide a lot of insight. Second comes a numerical technique of wide applicability.

Put both of these techniques into your problem-solving toolkit. You'll find them valuable not only in quantum mechanics, but whenever you need to solve a second-order ordinary differential equation.

5.1 Sketching energy eigenfunctions

Since this chapter is more mathematical than physical in character, I start off by writing the energy eigenequation (4.31) in the mathematically suggestive form

$$\frac{d^2\eta(x)}{dx^2} = -\frac{2m}{\hbar^2}[E - V(x)]\eta(x) = -\frac{2m}{\hbar^2}K_c(x)\eta(x) \tag{5.2}$$

which defines the "classical kinetic energy function" $K_c(x)$. This parallels the potential energy function: $V(x)$ is the potential energy that the classical system would have if the particle were located at x. I'm not saying that the particle *is* classical nor that it *does have* a location; indeed a quantal particle might not have a location. But $V(x)$ is the potential energy that the system would have if it were classical with the particle located at point x. In the same way $K_c(x)$ is the kinetic energy that a classical particle would have if the particle were located at x and total energy were E. Whereas no classical particle can ever have a negative kinetic energy, it is perfectly permissible for the classical kinetic energy function to be negative: in the graph that follows, $K_c(x)$ is negative on the left, positive in the center, and strongly negative on the right.

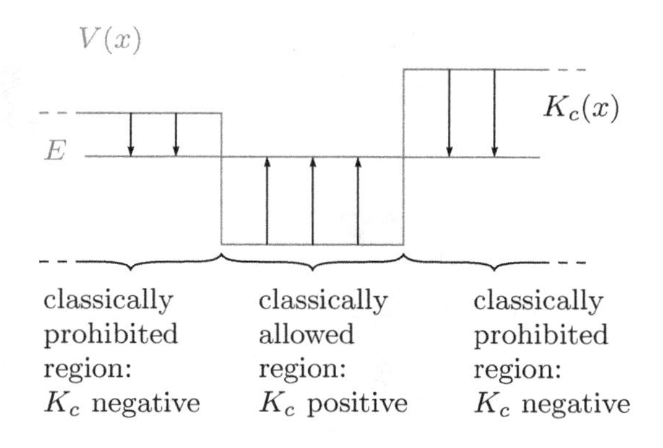

A region were $K_c(x)$ is positive or zero is called a "classically allowed region"; otherwise it is a "classically prohibited region".

Remember that

$$\frac{d\eta}{dx} \text{ represents slope;} \qquad \frac{d^2\eta}{dx^2} \text{ represents curvature.}$$

When curvature is positive, the slope increases as x increases (e.g. from negative to positive, or from positive small to positive large). When curvature is negative, the slope decreases as x increases.

Start off by thinking of a **classically allowed region** where $K_c(x)$ is constant and positive. Equation (5.2) says that if $\eta(x)$ is positive, then the curvature is negative, whereas if $\eta(x)$ is negative, then the curvature is positive. Furthermore, the size of the curvature depends on the size of $\eta(x)$:

when $\eta(x)$ is...	curvature is...
strongly positive	strongly negative
weakly positive	weakly negative
zero	zero
weakly negative	weakly positive
strongly negative	strongly positive

These observations allow us to find the character of $\eta(x)$ without finding a formal solution. If at one point $\eta(x)$ is positive with positive slope, then moving to the right $\eta(x)$ will grow because of the positive slope, but that growth rate will decline because of the negative curvature. Eventually the slope becomes zero and then negative, but the curvature continues negative. Because of the negative slope, $\eta(x)$ eventually plunges through $\eta(x) = 0$ (where its curvature is zero) and into regions where $\eta(x)$ is negative and hence the curvature is positive. The process repeats to produce the following graph:

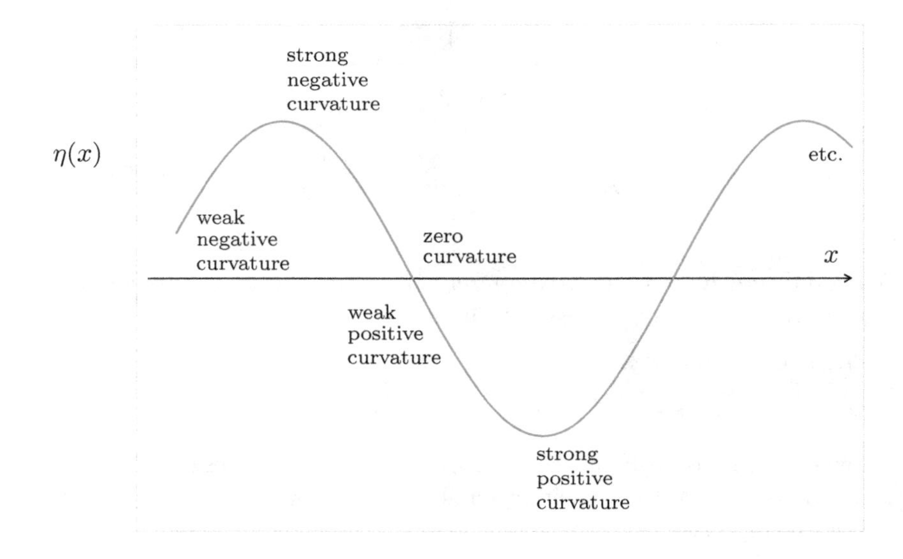

⟦You can solve differential equation (5.2) formally to obtain

$$\eta(x) = A\sin((\sqrt{2mK_c}/\hbar)x + \phi) \tag{5.3}$$

where A and ϕ are adjusted to fit the initial or boundary conditions. In fact, this is exactly the equation that we already solved at (4.17). The formal approach has the advantage of finding an exact expression for the wavelength. The informal approach has the advantage of building your intuition.⟧

The direct way of keeping track of curvature in this classically allowed region is

negative curvature when $\eta(x)$ is positive;
positive curvature when $\eta(x)$ is negative.

But this is sort of clunky: to keep track of curvature, you have to keep track of height. A compact way of keeping track of the signs is that

in a classically allowed region, (5.4)
the eigenfunction curves toward the axis.

It doesn't *slope* toward the axis, as you can see from the graph, it *curves* toward the axis. Draw a tangent to the energy eigenfunction: in a classically allowed region, the eigenfunction will fall between that tangent line and the axis.

In fact, the informal approach uncovers more than just the oscillatory character of $\eta(x)$. Equation (5.2) shows that when K_c is large and positive, the "curving toward" impetus is strong; when K_c is small and positive, that impetus is weak. Thus when K_c is large, the wavefunction takes tight turns and snaps back toward the axis; when K_c is small, it lethargically bends back toward the axis. And sure enough the formal approach at equation (5.3) shows that the wavelength λ depends on K_c through

$$\lambda = \frac{2\pi\hbar}{\sqrt{2mK_c}}, \tag{5.5}$$

so a large K_c results in a short wavelength — a "tight turn" toward the axis.

Now turn your attention to a **classically prohibited region** where $K_c(x)$ is constant and negative. Equation (5.2) says that if $\eta(x)$ is positive, then the curvature is positive. Once again we can uncover the character of $\eta(x)$ without finding a formal solution. If at one point $\eta(x)$ is positive with positive slope, then moving to the right $\eta(x)$ will grow because of the positive slope, and that growth rate increases because of the positive curvature. The slope becomes larger and larger and $\eta(x)$ rockets to infinity. Or, if $\eta(x)$ starts out negative with negative slope, then it rockets down to negative infinity. Or, if $\eta(x)$ starts out positive with negative slope, it might cross the axis before rocketing down to negative infinity, or it might dip down toward the axis without crossing it, before rocketing up to positive infinity.

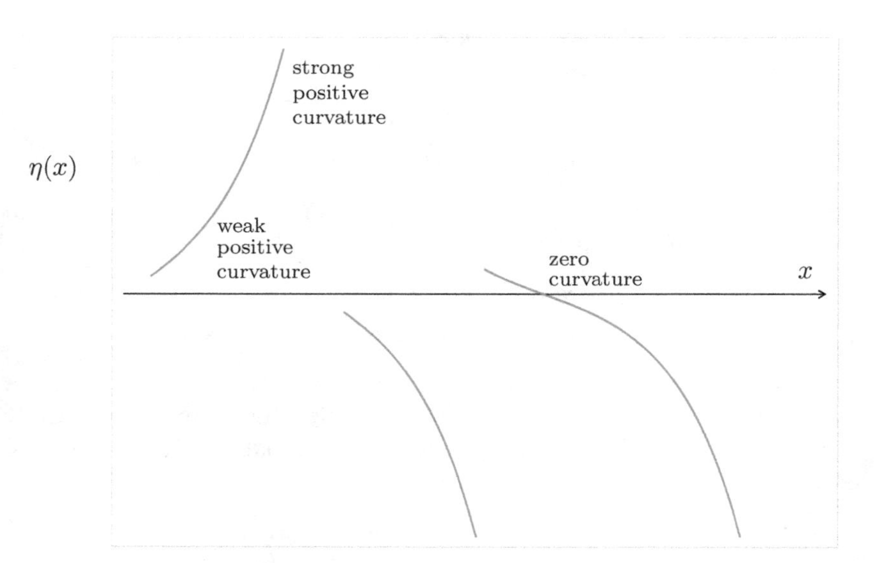

⟦You can solve differential equation (5.2) formally to obtain

$$\eta(x) = Ae^{+(\sqrt{2m|K_c|}/\hbar)x} + Be^{-(\sqrt{2m|K_c|}/\hbar)x}$$

where A and B are adjusted to fit the initial or boundary conditions.⟧

The direct way of keeping track of curvature in this classically prohibited region is

positive curvature when $\eta(x)$ is positive;
negative curvature when $\eta(x)$ is negative.

But a compact way is remembering that

in a classically prohibited region, (5.6)
the eigenfunction curves away from the axis.

Draw a tangent to the energy eigenfunction: in a classically prohibited region, that tangent line will fall between the eigenfunction and the axis.

Let's apply all these ideas to finding the character of energy eigenfunctions in a **finite square well**. Solve differential equation (5.2) for an energy E just above the bottom of the well. (I will draw the potential energy function in olive green, the energy E in blue, and the solution $\eta(x)$ in red.)

Suppose the wavefunction starts out on the left small and just above the axis. The region is strongly prohibited, that is $K_c(x)$ is strongly negative, so $\eta(x)$ curves strongly away from the axis. Then (at the dashed vertical line) the solution moves into a classically allowed region. But $K_c(x)$ is only weakly positive, so $\eta(x)$ curves only weakly toward the axis. By the time the solution gets to the right-hand classically prohibited region at the next dashed vertical line, $\eta(x)$ has only a weakly negative slope. In the prohibited region the slope increases as $\eta(x)$ curves strongly away from the axis and rockets off to infinity.

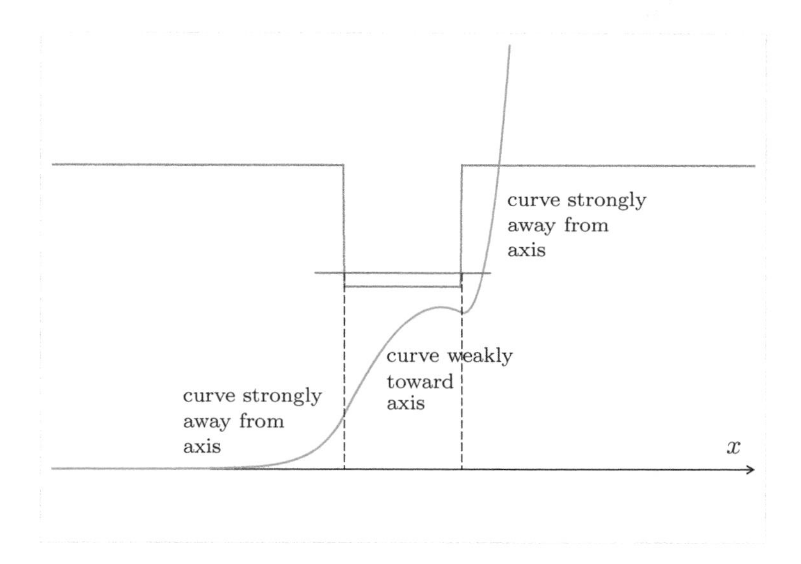

You should check that the curvatures and tangents of this energy eigenfunction strictly obey the rules set down at (5.4) and (5.6). What happens when $\eta(x)$ crosses a dashed vertical line, the boundary between a classically prohibited and a classically allowed region?

If you have studied differential equations you know that for any value of E, equation (5.2) has two linearly independent solutions. We've just sketched one of them. The other is the mirror image of it: small to the right and rocketing to infinity toward the left. Because of the "rocketing off to infinity" neither solution is normalizable. So these two solutions don't correspond to any physical energy eigenstate. To find such a solution we have to try a different energy.

So we try an energy slightly higher. Now the region on the left is not so strongly prohibited as it was before, so $\eta(x)$ curves away from the axis less dramatically. Then when it reaches the classically allowed region it curves more sharply toward the axis, so that it's strongly sloping downward when it reaches the right-hand prohibited region. But not strongly enough: it curves away from the axis and again rockets off to infinity — although this time not so dramatically.

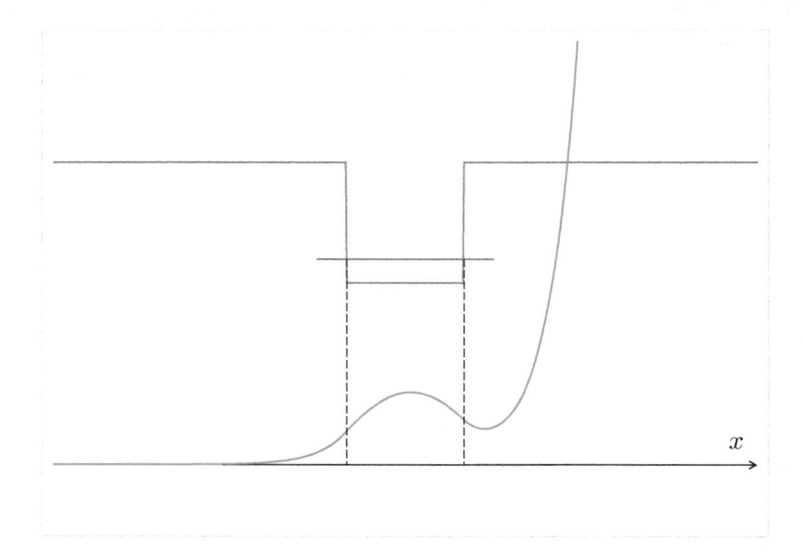

Once again we find a solution (and its mirror image is also a solution), but it's a non-physical, unnormalizable solution.

As we try energies higher and higher, the "rocketing to infinity" happens further and further to the right, until at one special energy it doesn't happen at all. Now the wavefunction *is* normalizable, and now we have found an energy eigenfunction.

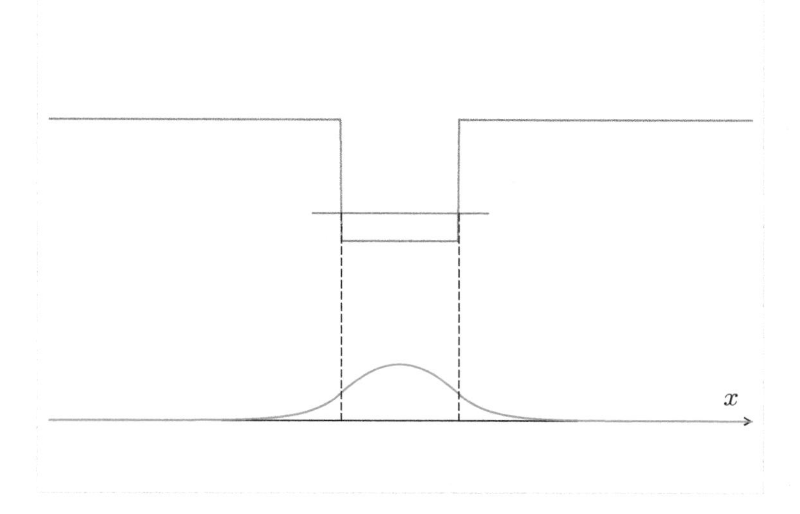

What happens when we try an energy slightly higher still? At the right-hand side the wavefunction now rockets off to *negative* infinity! With increased energies, the wavefunction rockets down to negative infinity with increased drama. But then at some point, the drama decreases: as the energy rises the wavefunction continues to go to negative infinity, but it does so more and more slowly. Finally at one special energy the wavefunction settles down exactly to zero as $x \to \infty$, and we've found a second energy eigenfunction.

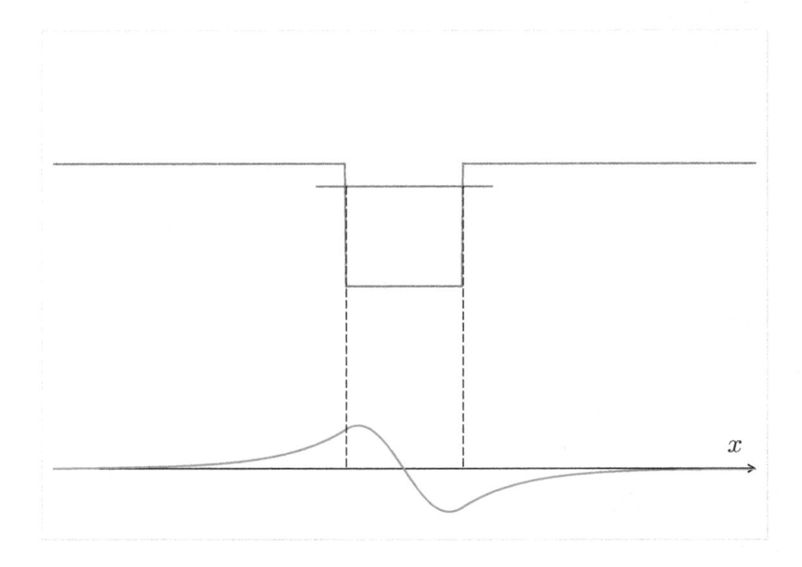

(The misconception concerning "pointlike particles moving across a node", discussed on page 140, applies to this state as well.)

The process continues: with still higher values of E, the wavefunction $\eta(x)$ diverges to positive infinity as $x \to \infty$ until we reach a third special energy eigenvalue, then to negative infinity until we reach a fourth. Higher and higher energies result in higher and higher values of K_c and hence stronger and stronger snaps back toward the axis. The first (lowest) eigenfunction has no nodes, the second has one node, the third will have two nodes, and in general the nth energy eigenfunction will have $n - 1$ nodes. (See also the discussion on page 190.)

Notice that for a potential energy function symmetric about a point, the energy eigenfunction is *either* symmetric or antisymmetric about that point. The energy eigenfunction does *not* need to possess the same symmetry as the potential energy function. (See also problem 5.7, "Parity".)

What about a **"lopsided" square well** that lacks symmetry? In the case sketched below the energy is strongly prohibited to the left, weakly prohibited to the right. Hence the wavefunction curves away sharply to the left, mildly to the right. The consequence is that the tail is short on the left, long on the right.

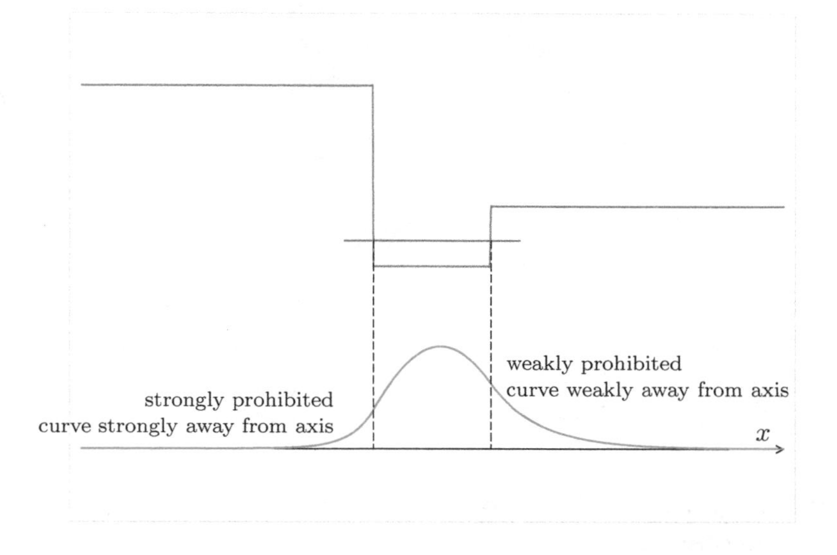

In some way it makes sense that the wavefunction tail should be longer where the classical prohibition is milder.

Now try a **square well with two different floor levels:**

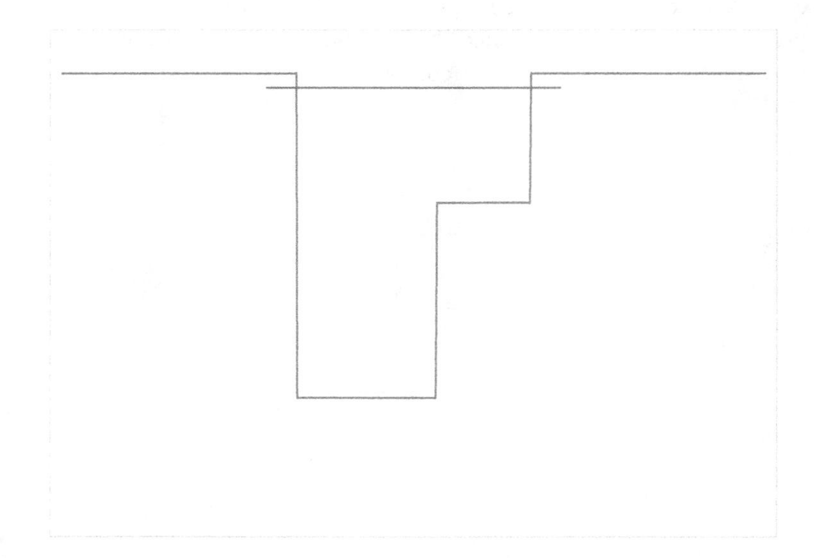

Within the deep left side of the well, K_c is relatively high, so the tendency for η to curve toward the axis is strong; within the shallow right side K_c is relatively low, so the tendency to curve toward the axis is weak. Thus within the deep side of the well, $\eta(x)$ snaps back toward the axis, taking the curves like an expertly driven sports car; within the shallow side $\eta(x)$ leisurely curves back toward the axis, curving like a student driver in a station wagon. Within the deep side, wavelength will be short and amplitude will be small; within the shallow side, wavelength will be longer and amplitude will be large (or at least the same size). One finds smaller amplitude at the deeper side of the well, and hence, all other things being equal, smaller probability for the particle to be in the deep side of the well.

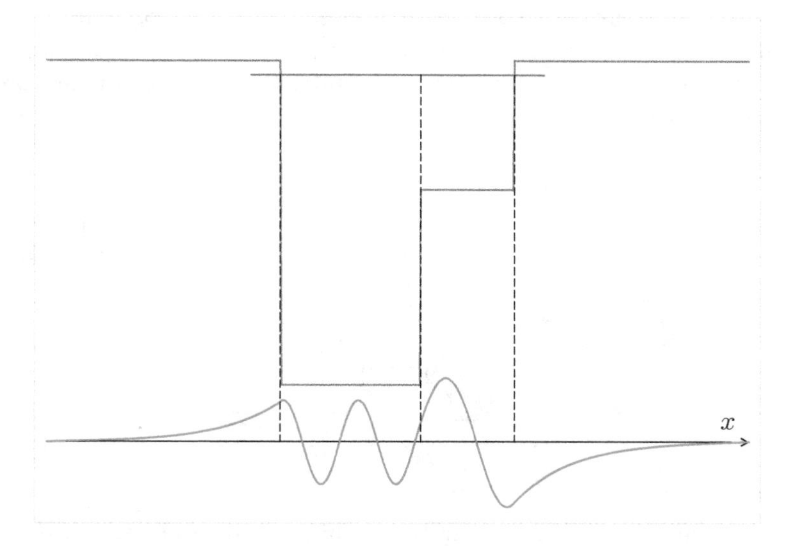

This might seem counterintuitive: Shouldn't it be more probable for the particle to be in the deep side? After all, if you throw a classical marble into a bowl it comes to rest at the deepest point and spends most of its time there. The problem with this analogy is that it compares a classical marble rolling with friction to a quantal situation without friction. Imagine a classical marble rolling instead in a frictionless bowl: it never does come to rest at the deepest point of the bowl. In fact, at the deepest point it moves fastest: the marble spends little time at the deepest point and a lot of time near the edges, where it moves slowly. The classical and quantal pictures don't correspond exactly (there's no such thing as an energy eigenstate in classical mechanics, the classical marble always has a position, and its description never has a node), but the two pictures agree that the particle has high probability of appearing where the potential energy function is shallow, not deep.

Similar results hold for three-level square wells, for four-level square wells, and so forth. And because any potential energy function can be approximated by a series of steps, similar results hold for any potential energy function.

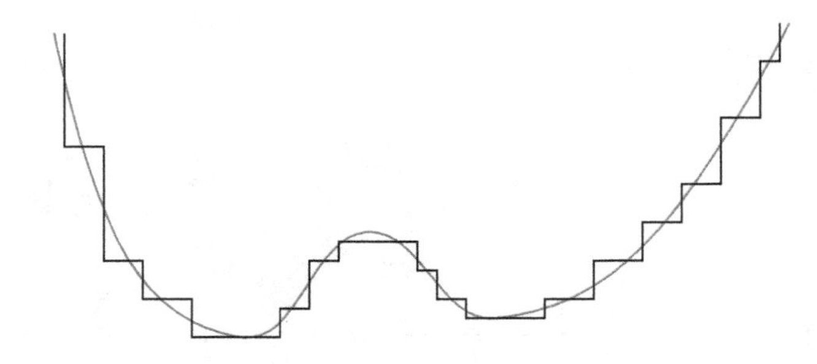

Number of nodes. For the infinite square well, the energy eigenfunction $\eta_n(x)$ has $n-1$ interior nodes. The following argument[1] shows that same holds for any one-dimensional potential energy function $V(x)$. Imagine a modified potential

$$V_a(x) = \begin{cases} \infty & x \leq -a \\ V(x) & -a < x < +a \\ \infty & +a \leq x \end{cases}.$$

When a is very small this is virtually an infinite square well, whose energy eigenfunctions we know. As a grows larger and larger, this potential becomes more and more like the potential of interest $V(x)$. During this expansion, can an extra node pop into an energy eigenfunction? If it does, then at the point x_p where it pops in the wavefunction vanishes, $\eta(x_p) = 0$, and its slope vanishes, $\eta'(x_p) = 0$. But the energy eigenproblem is a second-order ordinary differential equation: the only solution with $\eta(x_p) = 0$ and $\eta'(x_p) = 0$ is $\eta(x) = 0$ everywhere. This is not an eigenfunction. This can never happen.

[1]M. Moriconi, "Nodes of wavefunctions" *American Journal of Physics* **75** (March 2007) 284–285.

Summary

In classically prohibited regions, the eigenfunction magnitude declines while stepping away from the well: the stronger the prohibition, the more rapid the decline.

In classically allowed regions, the eigenfunction oscillates: in regions that are classically fast, the oscillation has small amplitude and short wavelength; in regions that are classically slow, the oscillation has large amplitude and long wavelength.

If the potential energy function is symmetric under reflection about a point, the eigenfunction will be either symmetric or antisymmetric under the same reflection.

The nth energy eigenfunction has $n - 1$ nodes.

Quantum mechanics involves situations (very small) and phenomena (interference, entanglement) remote from daily experience. And the energy eigenproblem, so central to quantum mechanics, does not arise in classical mechanics at all. Some people conclude from these facts that one cannot develop intuition about quantum mechanics, but that is false: the techniques of this section *do* allow you to develop a feel for the character of energy eigenstates. Just as chess playing or figure skating must be studied and practiced to develop proficiency, so quantum mechanics must be studied and practiced to develop intuition. If people don't develop intuition regarding quantum mechanics, it's not because quantum mechanics is intrinsically fantastic; it's because these people never try.

Problems

5.1 Would you buy a used eigenfunction from this man?
(recommended problem)
The four drawings below and on the next pages show four one-dimensional potential energy functions $V(x)$ (in olive green) along with candidate energy eigenfunctions $\eta(x)$ (in red) that purport to associate with those potential energy functions. There is something wrong with every candidate. Using the letter codes below, identify all eigenfunction errors, *and* sketch a qualitatively correct eigenfunction for each potential.

The energy eigenfunction is drawn incorrectly because:

A. Wrong curvature. (It curves toward the axis in a classically prohibited region or away from the axis in a classically allowed region.)
B. Its wavy part has the wrong number of nodes.
C. The amplitude of the wavy part varies incorrectly.
D. The wavelength of the wavy part varies incorrectly.
E. One or more of the declining tails has the wrong length.

a.

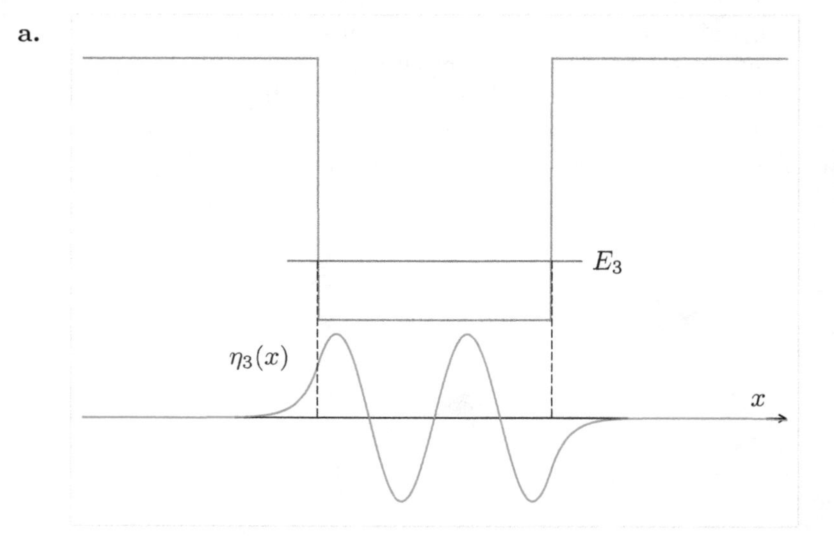

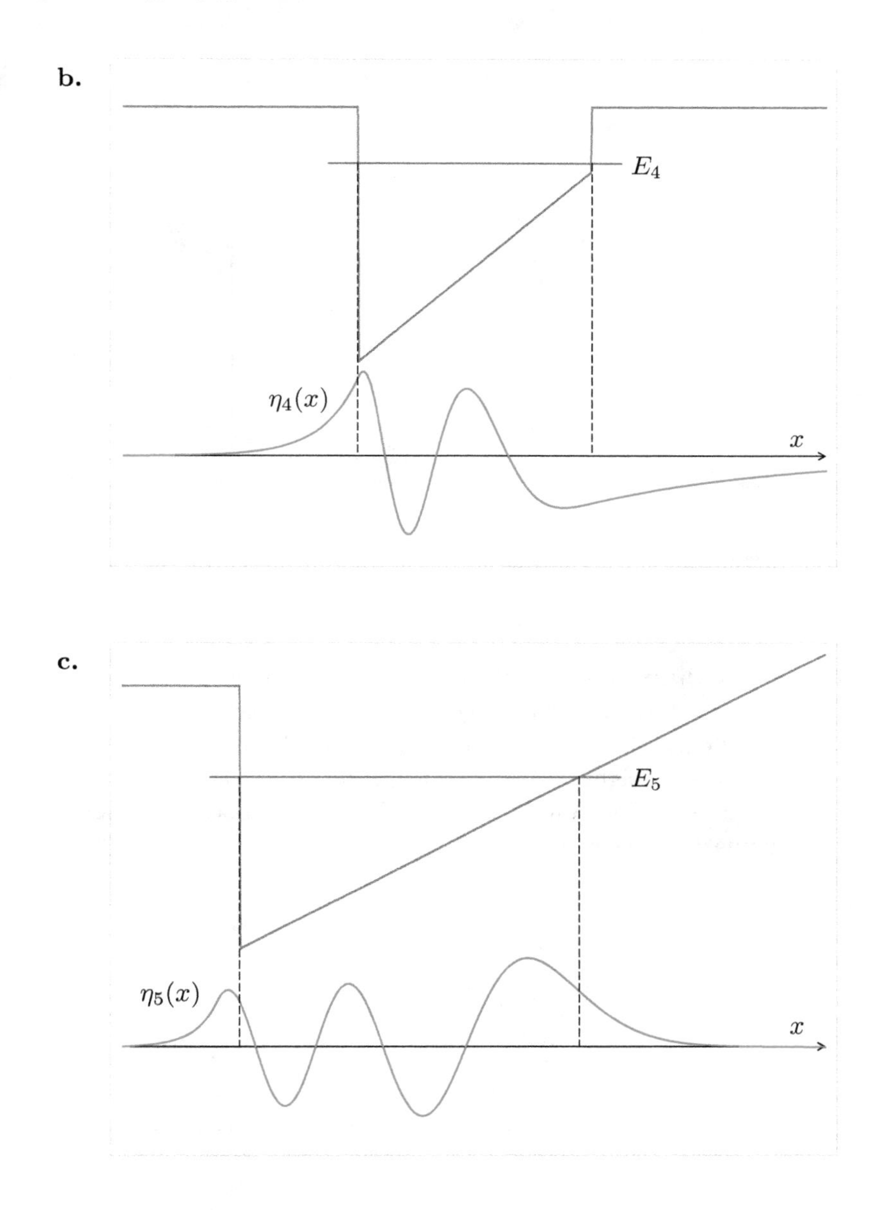
b.
E_4
$\eta_4(x)$
x
c.
E_5
$\eta_5(x)$
x

d.

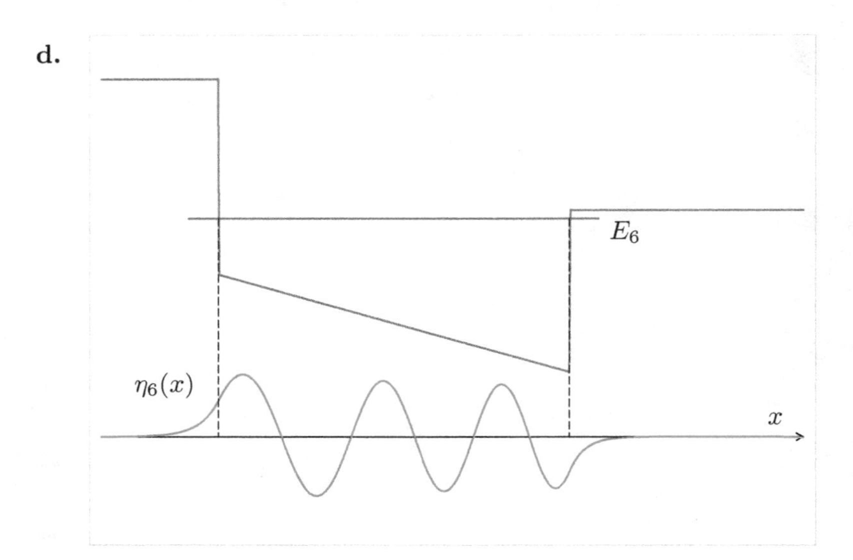

5.2 **Simple harmonic oscillator energy eigenfunctions**

Here are sketches of the three lowest-energy eigenfunctions for the potential energy function $V(x) = \frac{1}{2}kx^2$ (called the "simple harmonic oscillator"). In eight sentences or fewer, describe how these energy eigenfunctions do (or don't!) display the characteristics discussed in the summary on page 191.

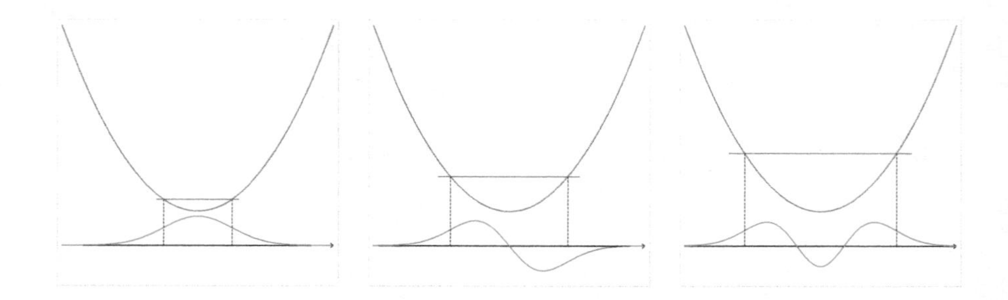

5.3 **de Broglie wavelength**
Compare equation (5.5) to the formula for de Broglie wavelength. Does this shed any light on the question (see page 46) of "what is waving" in a de Broglie wave?

5.4 **Wavelength as a function of K_c**
Before equation (5.5) we provided an informal argument that the wavelength λ would decrease with increasing K_c. This argument didn't say whether λ would vary as $1/K_c$, or as $1/\sqrt{K_c}$, or even as $e^{-K_c/(\text{constant})}$. Produce a dimensional argument showing that if λ depends only on $\hbar$, m, and K_c, then it must vary as $\hbar/\sqrt{mK_c}$.

5.5 **"At least the same size amplitude"**
Page 188 claims that in the two-level square well, the amplitude of $\eta(x)$ on the right would be larger "or at least the same size" as the amplitude on the left. Under what conditions will the amplitude be the same size?

5.6 **Placement of nodes**

Let $\eta_n(x)$ and $\eta_m(x)$ be solutions to

$$-\frac{\hbar^2}{2M}\eta''_m(x) + V(x)\eta_m(x) = E_m\eta_m(x) \tag{5.7}$$

$$-\frac{\hbar^2}{2M}\eta''_n(x) + V(x)\eta_n(x) = E_n\eta_n(x) \tag{5.8}$$

with $E_m > E_n$. The Sturm comparison theorem states that between any two nodes of $\eta_n(x)$ there exists at least one node of $\eta_m(x)$. Prove the theorem through contradiction by following these steps:

a. Multiply (5.7) by η_n, multiply (5.8) by η_m, and subtract to show that

$$-\frac{\hbar^2}{2M}[\eta'_m(x)\eta_n(x) - \eta_m(x)\eta'_n(x)]' = (E_m - E_n)\eta_m(x)\eta_n(x). \tag{5.9}$$

b. Call two adjacent nodes of $\eta_n(x)$ by the names x_1 and x_2. Argue that we can select $\eta_n(x)$ to be always positive for $x_1 < x < x_2$, and show that with this selection $\eta'_n(x_1) > 0$ while $\eta'_n(x_2) < 0$.

c. Integrate equation (5.9) from x_1 to x_2, producing

$$-\frac{\hbar^2}{2M}[-\eta_m(x_2)\eta'_n(x_2) + \eta_m(x_1)\eta'_n(x_1)]$$
$$= (E_m - E_n)\int_{x_1}^{x_2} \eta_m(x)\eta_n(x)\,dx. \tag{5.10}$$

d. If $\eta_m(x)$ does *not* have a zero within $x_1 < x < x_2$, then argue that we can select $\eta_m(x)$ always positive on the same interval, including the endpoints.

The assumption that "$\eta_m(x)$ does *not* have a zero" hence implies that the left-hand side of (5.10) is strictly negative, while the right-hand side is strictly positive. This assumption, therefore, must be false.

5.7 **Parity**

a. Think of an arbitrary potential energy function $V(x)$. Now think of its mirror image potential energy function $U(x) = V(-x)$ Show that if $\eta(x)$ is an eigenfunction of $V(x)$ with energy E, then $\sigma(x) = \eta(-x)$ is an eigenfunction of $U(x)$ with the same energy.

b. If $V(x)$ is symmetric under reflection about the origin, that is $U(x) = V(x)$, you might think that $\sigma(x) = \eta(x)$. But no! This identification ignores global phase freedom (pages 107 and 124). Show that in fact $\sigma(x) = r\eta(x)$ where the "overall phase factor" r is a complex number with magnitude 1.

c. The overall phase factor r is a number, not a function of x: the same phase factor r applies at $x = 2$ ($\eta(-2) = r\eta(2)$), at $x = 7$ ($\eta(-7) = r\eta(7)$), and at $x = -2$ ($\eta(2) = r\eta(-2)$). Conclude that r can't be any old complex number with magnitude 1, it must be either $+1$ or -1.

Energy eigenfunctions symmetric under reflection, $\eta(x) = \eta(-x)$, are said to have "even parity" while those antisymmetric under reflection, $\eta(x) = -\eta(-x)$, are said to have "odd parity".

5.8 **Scaling**

Think of an arbitrary potential energy function $V(x)$, for example perhaps the one sketched on the left below. Now think of another potential energy function $U(y)$ that is half the width and four times the depth/height of $V(x)$, namely $U(y) = 4V(x)$ where $y = x/2$. Without solving the energy eigenproblem for either $V(x)$ or $U(y)$, I want to find how the energy eigenvalues of $U(y)$ relate to those of $V(x)$.

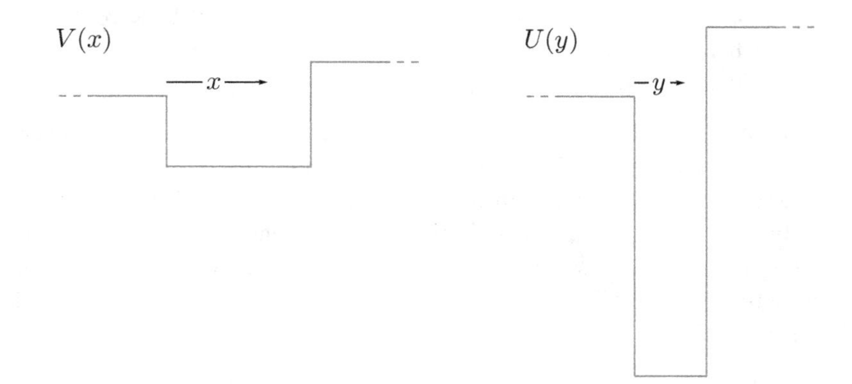

Show that if $\eta(x)$ is an eigenfunction of $V(x)$ with energy E, then $\sigma(y) = \eta(x)$ is an eigenfunction of $U(y)$. What is the corresponding energy? After working this problem for the scale factor 2, repeat for a general scale factor s so that $U(y) = s^2V(x)$ where $y = x/s$.

⟦This problem has a different cast from most: instead of giving you a problem and asking you to solve it, I'm asking you to find the relationship between the solutions of two different problems, neither of which you've solved. My thesis adviser, Michael Fisher, called this "Juicing an orange without breaking its peel."⟧

5.2 Scaled quantities

Look again at the quantal energy eigenproblem (5.2)

$$\frac{d^2\eta(x)}{dx^2} = -\frac{2m}{\hbar^2}[E - V(x)]\eta(x). \tag{5.11}$$

Suppose you want to write a computer program to solve this problem for the lopsided square well with potential energy function

$$V(x) = \begin{cases} V_1 & x < 0 \\ 0 & 0 < x < L \\ V_2 & L < x \end{cases}. \tag{5.12}$$

The program would have to take as input the particle mass m, the energy E, the potential well length L, and the potential energy values V_1 and V_2. Five parameters! Once the program is written, you'd have to spend a lot of time typing in these parameters and exploring the five-dimensional parameter space to find interesting values. Furthermore, these parameters have inconvenient magnitudes like the electron's mass 9.11×10^{-31} kg or the length of a typical carbon nanotube 1.41×10^{-10} m. Isn't there an easier way to set up this problem?

There is. The characteristic length for this problem is L. If you try to combine the parameters L, m, and $\hbar$ to form a quantity with the dimensions of energy (see sample problem 5.2.1 on page 200) you will find that there is only one way: this problem's characteristic energy is $E_c = \hbar^2/mL^2$. Define the dimensionless length variable $\tilde{x} = x/L$, the dimensionless energy parameter $\tilde{E} = E/E_c$, and the dimensionless potential energy function $\tilde{V}(\tilde{x}) = V(\tilde{x}L)/E_c = V(x)/E_c$.

In terms of these new so-called "scaled quantities" the quantal energy eigenproblem is

$$\frac{d^2\eta(\tilde{x})}{d\tilde{x}^2}\frac{1}{L^2} = -\frac{2m}{\hbar^2}\left[\frac{\hbar^2}{mL^2}\right][\tilde{E} - \tilde{V}(\tilde{x})]\eta(\tilde{x})$$

or

$$\frac{d^2\eta(\tilde{x})}{d\tilde{x}^2} = -2[\tilde{E} - \tilde{V}(\tilde{x})]\eta(\tilde{x}) \tag{5.13}$$

where

$$\tilde{V}(\tilde{x}) = \begin{cases} \tilde{V}_1 & \tilde{x} < 0 \\ 0 & 0 < \tilde{x} < 1 \\ \tilde{V}_2 & 1 < \tilde{x} \end{cases}. \tag{5.14}$$

The scaled problem has many advantages. Instead of five there are only three parameters: $\tilde{E}$, $\tilde{V}_1$, and $\tilde{V}_2$. And those parameters have nicely sized values like 1 or 0.5 or 6. But it has the disadvantage that you have to write down all those tildes. Because no one likes to write down tildes, we just drop them, writing the problem as

$$\frac{d^2\eta(x)}{dx^2} = -2[E - V(x)]\eta(x) \tag{5.15}$$

where

$$V(x) = \begin{cases} V_1 & x < 0 \\ 0 & 0 < x < 1 \\ V_2 & 1 < x \end{cases} \tag{5.16}$$

and saying that these equations are written down "using scaled quantities".

When you compare these equations with equations (5.11) and (5.12), you see that we would get the same result if we had simply said "let $\hbar = m = L = 1$". This phrase as stated is of course absurd: $\hbar$ is *not* equal to 1; $\hbar$, m, and L *do* have dimensions. But some people don't like to explain what they're doing so they do say this as shorthand. Whenever you hear this phrase, remember that it covers up a more elaborate — and more interesting — truth.

5.2.1 *Sample Problem: Characteristic energy*

Show that there is only one way to combine the quantities L, m, and $\hbar$ to form a quantity with the dimensions of energy, and find an expression for this so-called characteristic energy E_c.

Solution:

quantity	dimensions
L	[length]
m	[mass]
$\hbar$	[mass] × [length]^2/[time]
E_c	[mass] × [length]^2/[time]^2

If we are to build E_c out of L, m, and $\hbar$, we must start with $\hbar$, because that's the only source of the dimension [time]. And in fact we must start with $\hbar^2$, because that's the only way to make a [time]^2.

quantity	dimensions
L	[length]
m	[mass]
$\hbar^2$	[mass]^2 × [length]^4/[time]^2
E_c	[mass] × [length]^2/[time]^2

But $\hbar^2$ has too many factors of [mass] and [length] to make an energy. There is only one way to get rid of them: to divide by m once and by L twice.

quantity	dimensions
$\hbar^2/mL^2$	[mass] × [length]^2/[time]^2
E_c	[mass] × [length]^2/[time]^2

There is only one possible characteristic energy, and it is $E_c = \hbar^2/mL^2$.

Problems

5.9 Characteristic time

Find the characteristic time for the square well problem by combining the parameters L, m, and $\hbar$ to form a quantity with the dimensions of time. Compare this characteristic time to the infinite square well revival time found at equation (4.25).

5.10 Scaling for the simple harmonic oscillator

(recommended problem)

Execute the scaling strategy for the simple harmonic oscillator potential energy function $V(x) = \frac{1}{2}kx^2$. What is the characteristic length in terms of k, $\hbar$, and m? What is the resulting scaled energy eigenproblem? If you didn't like to explain what you were doing, how would you use shorthand to describe the result of this scaling strategy?

5.3 Numerical solution of the energy eigenproblem

Now that the quantities are scaled, we return to our task of writing a computer program to solve, numerically, the energy eigenproblem. In order to fit the potential energy function $V(x)$ and the energy eigenfunction $\eta(x)$ into a finite computer, we must of course approximate those continuum functions through their values on a finite grid. The grid points are separated by a small quantity Δ. It is straightforward to replace the function $V(x)$ with grid values V_i and the function $\eta(x)$ with grid values η_i. But what should we do with the second derivative $d^2\eta/dx^2$?

Start with a representation of the grid function η_i:

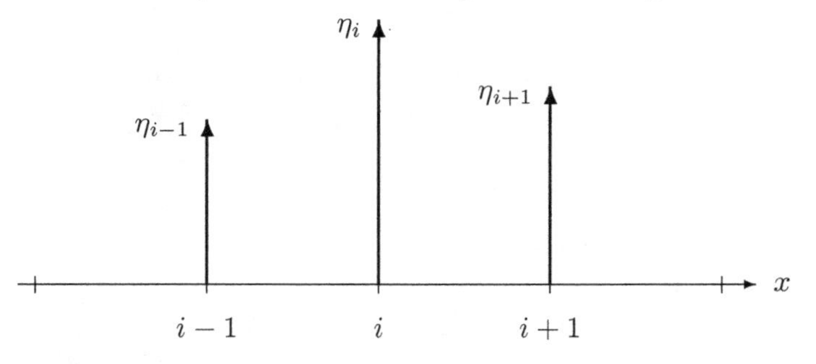

The slope at a point halfway between points $i-1$ and i (represented by the left dot in the figure below) is approximately

$$\frac{\eta_i - \eta_{i-1}}{\Delta},$$

while the slope half way between the points i and $i+1$ (represented by the right dot) is approximately

$$\frac{\eta_{i+1} - \eta_i}{\Delta}.$$

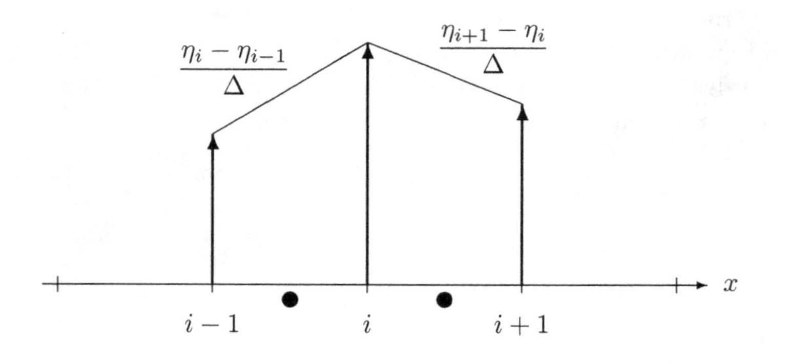

An approximation for the second derivative at point i is the change in slope divided by the change in distance

$$\frac{\dfrac{\eta_{i+1} - \eta_i}{\Delta} - \dfrac{\eta_i - \eta_{i-1}}{\Delta}}{\Delta}$$

so at point i we approximate

$$\frac{d^2\eta}{dx^2} \approx \frac{\eta_{i+1} - 2\eta_i + \eta_{i-1}}{\Delta^2}. \tag{5.17}$$

The discretized version of the energy eigenproblem (5.15) is thus

$$\frac{\eta_{i+1} - 2\eta_i + \eta_{i-1}}{\Delta^2} = -2[E - V_i]\eta_i \tag{5.18}$$

which rearranges to

$$\eta_{i+1} = 2[1 + \Delta^2(V_i - E)]\eta_i - \eta_{i-1}. \tag{5.19}$$

The algorithm then proceeds from left to right. Start in a classically prohibited region and select $\eta_1 = 0$, $\eta_2 = 0.001$. Then find

$$\eta_3 = 2[1 + \Delta^2(V_2 - E)]\eta_2 - \eta_1.$$

Now that you know η_3, find

$$\eta_4 = 2[1 + \Delta^2(V_3 - E)]\eta_3 - \eta_2.$$

Continue until you know η_i at every grid point.

For most values of E, this algorithm will result in a solution that rockets to $\pm\infty$ at the far right. When you pick a value of E where the solution approaches zero at the far right, you've found an energy eigenvalue. The algorithm is called "shooting", because it resembles shooting an arrow at a fixed target: your first shot might be too high, your second too low, so you try something between until you home in on your target.

Problems

5.11 **Program**

a. Implement the shooting algorithm using a computer spreadsheet, your favorite programming language, or in any other way. You will have to select reasonable values for Δ and η_2.
b. Check your implementation by solving the energy eigenproblem for a free particle and for an infinite square well.
c. Find the three lowest-energy eigenvalues for a square well with $V_1 = V_2 = 30$. Do the corresponding eigenfunctions have the qualitative character you expect?
d. Repeat for a square well with $V_1 = 50$ and $V_2 = 30$.

5.12 **Algorithm parameter**

Below equation (5.19) I suggested that you start the stepping algorithm with $\eta_1 = 0$, $\eta_2 = 0.001$. What would have happened had you selected $\eta_1 = 0$, $\eta_2 = 0.003$ instead?

5.13 **Simple harmonic oscillator**

(Work problem 5.10 on page 201 before working this one.)

Implement the algorithm for a simple harmonic oscillator using scaled quantities. Find the five lowest-energy eigenvalues, and compare them to the analytic results 0.5, 1.5, 2.5, 3.5, and 4.5.

5.14 **Questions** (recommended problem)

Update your list of quantum mechanics questions that you started at problem 1.17 on page 46. Write down new questions and, if you have uncovered answers to any of your old questions, write them down briefly.

⟦For example, one of my questions would be: "For any value of E — energy eigenvalue or no — equation (5.2) has two linearly independent solutions. We saw on page 183 that often the two linearly independent solutions are mirror images, one rocketing off to infinity as $x \to +\infty$ and the other rocketing off to infinity as $x \to -\infty$. But what about the energy eigenfunctions, which go to zero as $x \to \pm\infty$? What does the other linearly independent solution look like then?"⟧

Chapter 6

Identical Particles

6.1 Two or three identical particles

Please review section 4.4, "Wavefunction: Two particles in one or three dimensions", on page 127. In that section we talked about two different particles, say an electron and a neutron. We set up a grid, discussed bin amplitudes $\psi_{i,j}$, and talked about the limit as the width of each bin shrank to zero.

There is a parallel development for two identical particles, but with one twist. Here is the situation when one particle is found in bin 5, the other in bin 8:

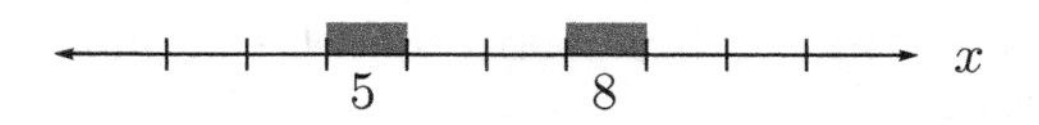

And here is the situation when one particle is found in bin 8, the other in bin 5:

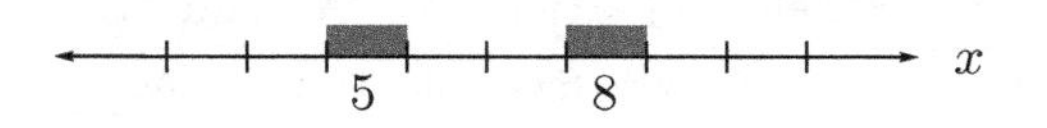

No difference, of course... that's the *meaning* of "identical". And of course this holds not only for bins 5 and 8, but for any pair of bins i and j, even if $i = j$. (If the two particles don't interact, it is perfectly plausible for both of them to occupy the same bin at the same time.)

What does this mean for the state of a system with two identical particles? Suppose that, by hook or by crook, we come up with a set of bin amplitudes $\psi_{i,j}$ that describes the state of the system. Then the set of amplitudes $\phi_{i,j} = \psi_{j,i}$ describes that state just as well as the original set $\psi_{i,j}$. Does this mean that $\phi_{i,j} = \psi_{i,j}$? Not at all. Remember global phase freedom (pages 107 and 124): If every bin amplitude is multiplied by the same "overall phase factor" — a complex number with magnitude unity — then the resulting set of amplitudes describes the state just as well as the original set did. Calling that overall phase factor s, we conclude that $\phi_{i,j} = s\psi_{i,j}$.

But, because $\phi_{i,j} = \psi_{j,i}$, the original set of amplitudes must satisfy $\psi_{j,i} = s\psi_{i,j}$. The variable name s comes from "swap": when we swap subscripts, we introduce a factor of s. The quantity s is a number... not a function of i or j. For example, the same value of s must work for $\psi_{8,5} = s\psi_{5,8}$, for $\psi_{7,3} = s\psi_{3,7}$, for $\psi_{5,8} = s\psi_{8,5}$, ... Wait. What was that last one? Put together the first and last examples:

$$\psi_{8,5} = s\psi_{5,8} = s(s\psi_{8,5}) = s^2\psi_{8,5}.$$

Clearly, $s^2 = 1$, so s can't be any old complex number with magnitude unity: it can be only $s = +1$ or $s = -1$.

Execute the now-familiar program of turning bin amplitudes into amplitude density, that is wavefunction, to find that

$$\psi(x_A, x_B) = +\psi(x_B, x_A) \quad \text{or} \quad \psi(x_A, x_B) = -\psi(x_B, x_A). \tag{6.1}$$

The first kind of wavefunction is called "symmetric under coordinate swapping", the second is called "antisymmetric under coordinate swapping". This requirement for symmetry or antisymmetry under coordinate swapping is called the Pauli[1] principle.

It might distress you to see variables like x_A: doesn't x_A mean the position of particle "A" while x_B means the position of particle "B"? So doesn't this terminology label the particles as "A" and "B", which would

[1]Wolfgang Pauli (1900–1958), Vienna-born Swiss physicist, was one of the founders of quantum mechanics. In 1924 he proposed the "exclusion principle", ancestor of today's symmetry/antisymmetry requirement; in 1926 he produced the first solution for the energy eigenproblem for atomic hydrogen; in 1930 he proposed the existence of the neutrino, a prediction confirmed experimentally in 1956; in 1934 he and "Viki" Weisskopf discovered how to make sense of relativistic quantum mechanics by realizing that the solutions to relativistic quantal equations do not give an amplitude for a single particle to have a position (technically, a wavefunction), but rather an amplitude for an additional particle to be created at a position or for an existing particle to be annihilated at a position (technically, a creation or annihilation operator). He originated the insult, applied to ideas that cannot be tested, that they are "not even wrong".

violate our initial requirement that the particles be identical? The answer is that this terminology does *not* label one particle "A" and the other particle "B". Instead, it labels one *point* "A" and the other *point* "B". Look back to the figures on page 205: the numbers 5 and 8 label *bins*, not *particles*, so when these bins shrink to zero the variables x_A and x_B apply to *points*, not *particles*. That's why I like to call these wavefunctions "(anti)symmetric under swap of coordinates". But you'll hear people using terms like "(anti)symmetric under particle swapping" or "...under particle interchange" or "...under particle exchange".

What if the two particles are in three-dimensional space, and what if they have spin? In that case, the swap applies to *all* the coordinates: using the sans serif notation of equation (4.106),

$$\psi(\mathsf{x}_A, \mathsf{x}_B) = +\psi(\mathsf{x}_B, \mathsf{x}_A) \quad \text{or} \quad \psi(\mathsf{x}_A, \mathsf{x}_B) = -\psi(\mathsf{x}_B, \mathsf{x}_A). \tag{6.2}$$

From our argument so far, two identical electrons might be in a symmetric or an antisymmetric state. Similarly for two identical neutrons, two identical silver atoms, etc. But it's an empirical fact that the swap symmetry depends only on the kind of particle involved: Two electrons are always antisymmetric under swapping. Two ^{4}He atoms (both in their ground state) are always symmetric. Particles that are always symmetric under swapping are called bosons;[2] those always antisymmetric under swapping are called fermions.[3]

[2]Named for Satyendra Bose (1894–1974) of India. He contributed to fields ranging from chemistry to school administration, but his signal contribution was elucidating the statistics of photons. Remarkably, he made this discovery in 1922, three years before Schrödinger developed the concept of wavefunction.

[3]Named for Enrico Fermi. See footnote on page 170.

What about three particles? The wavefunction of three identical bosons must be completely symmetric, that is, symmetric under swaps of any coordinate pair:

$$\begin{aligned} &+\psi(x_A, x_B, x_C) \\ &= +\psi(x_A, x_C, x_B) \\ &= +\psi(x_C, x_A, x_B) \\ &= +\psi(x_C, x_B, x_A) \\ &= +\psi(x_B, x_C, x_A) \\ &= +\psi(x_B, x_A, x_C). \end{aligned} \tag{6.3}$$

(These $6 = 3!$ permutations are listed in the sequence called "plain changes" or "the Johnson-Trotter sequence". This sequence has the admirable property that each permutation differs from its predecessor by a single swap of adjacent letters.[4]) Whereas the wavefunction of three identical fermions must be completely antisymmetric, that is, antisymmetric under swaps of any coordinate pair:

$$\begin{aligned} &+\psi(x_A, x_B, x_C) \\ &= -\psi(x_A, x_C, x_B) \\ &= +\psi(x_C, x_A, x_B) \\ &= -\psi(x_C, x_B, x_A) \\ &= +\psi(x_B, x_C, x_A) \\ &= -\psi(x_B, x_A, x_C). \end{aligned} \tag{6.4}$$

As you would expect, Pauli's requirement of complete symmetry or antisymmetry under swaps of any coordinate pair holds for any number of identical particles.

6.2 Symmetrization and antisymmetrization

Given the importance of wavefunctions symmetric or antisymmetric under coordinate swaps, it makes sense to investigate the mathematics of such "permutation symmetry". This section treats systems of two or three

[4]Donald Knuth, *The Art of Computer Programming*, volume 4A, "Combinatorial Algorithms, Part 1" (Addison-Wesley, Boston, 1997) section 7.2.1.2, "Generating all permutations".

particles; the generalization to systems of four or more particles is straightforward.

Start with any two-variable garden-variety function $f(\mathsf{x}_A, \mathsf{x}_B)$, not necessarily symmetric or antisymmetric. Can that function be used as a "seed" to build a symmetric or antisymmetric function? It can. The function

$$s(\mathsf{x}_A, \mathsf{x}_B) = f(\mathsf{x}_A, \mathsf{x}_B) + f(\mathsf{x}_B, \mathsf{x}_A) \tag{6.5}$$

is symmetric under swapping while the function

$$a(\mathsf{x}_A, \mathsf{x}_B) = f(\mathsf{x}_A, \mathsf{x}_B) - f(\mathsf{x}_B, \mathsf{x}_A) \tag{6.6}$$

is antisymmetric. If you don't believe me, try it out:

$$\begin{aligned} s(5,2) &= f(5,2) + f(2,5) \\ s(2,5) &= f(2,5) + f(5,2) \end{aligned}$$

so clearly $s(5,2) = s(2,5)$. Meanwhile

$$\begin{aligned} a(5,2) &= f(5,2) - f(2,5) \\ a(2,5) &= f(2,5) - f(5,2) \end{aligned}$$

so just as clearly $a(5,2) = -a(2,5)$.

Can this be generalized to three variables? Start with a three-variable garden-variety function $f(\mathsf{x}_A, \mathsf{x}_B, \mathsf{x}_C)$. The function

$$\begin{aligned} s(\mathsf{x}_A, \mathsf{x}_B, \mathsf{x}_C) = \quad & f(\mathsf{x}_A, \mathsf{x}_B, \mathsf{x}_C) \\ & +f(\mathsf{x}_A, \mathsf{x}_C, \mathsf{x}_B) \\ & +f(\mathsf{x}_C, \mathsf{x}_A, \mathsf{x}_B) \\ & +f(\mathsf{x}_C, \mathsf{x}_B, \mathsf{x}_A) \\ & +f(\mathsf{x}_B, \mathsf{x}_C, \mathsf{x}_A) \\ & +f(\mathsf{x}_B, \mathsf{x}_A, \mathsf{x}_C) \end{aligned} \tag{6.7}$$

is completely symmetric while the function

$$\begin{aligned} a(\mathsf{x}_A, \mathsf{x}_B, \mathsf{x}_C) = \quad & f(\mathsf{x}_A, \mathsf{x}_B, \mathsf{x}_C) \\ & -f(\mathsf{x}_A, \mathsf{x}_C, \mathsf{x}_B) \\ & +f(\mathsf{x}_C, \mathsf{x}_A, \mathsf{x}_B) \\ & -f(\mathsf{x}_C, \mathsf{x}_B, \mathsf{x}_A) \\ & +f(\mathsf{x}_B, \mathsf{x}_C, \mathsf{x}_A) \\ & -f(\mathsf{x}_B, \mathsf{x}_A, \mathsf{x}_C) \end{aligned} \tag{6.8}$$

is completely antisymmetric. Once again, if you don't believe me I invite you to try it out with $x_A = 5$, $x_B = 2$, and $x_C = 7$.

This trick is often used when the seed function is a product,

$$f(x_A, x_B, x_C) = f_1(x_A) f_2(x_B) f_3(x_C), \tag{6.9}$$

in which case you may think of the symmetrization/antisymmetrization machinery as being the sum over all permutations of the coordinates x_A, x_B, and x_C, as above, or as the sum over all permutations of the functions $f_1(x)$, $f_2(x)$, and $f_3(x)$: the function

$$\begin{aligned} s(x_A, x_B, x_C) = \quad & f_1(x_A) f_2(x_B) f_3(x_C) \\ & + f_1(x_A) f_3(x_B) f_2(x_C) \\ & + f_3(x_A) f_1(x_B) f_2(x_C) \\ & + f_3(x_A) f_2(x_B) f_1(x_C) \\ & + f_2(x_A) f_3(x_B) f_1(x_C) \\ & + f_2(x_A) f_1(x_B) f_3(x_C) \end{aligned} \tag{6.10}$$

is completely symmetric while the function

$$\begin{aligned} a(x_A, x_B, x_C) = \quad & f_1(x_A) f_2(x_B) f_3(x_C) \\ & - f_1(x_A) f_3(x_B) f_2(x_C) \\ & + f_3(x_A) f_1(x_B) f_2(x_C) \\ & - f_3(x_A) f_2(x_B) f_1(x_C) \\ & + f_2(x_A) f_3(x_B) f_1(x_C) \\ & - f_2(x_A) f_1(x_B) f_3(x_C) \end{aligned} \tag{6.11}$$

is completely antisymmetric. Some people write this last expression as the determinant of a matrix

$$a(x_A, x_B, x_C) = \begin{vmatrix} f_1(x_A) & f_2(x_A) & f_3(x_A) \\ f_1(x_B) & f_2(x_B) & f_3(x_B) \\ f_1(x_C) & f_2(x_C) & f_3(x_C) \end{vmatrix}, \tag{6.12}$$

and call it the "Slater[5] determinant". I personally think this terminology confuses the issue (the expression works only if the seed function is a product of one-variable functions, it suppresses the delightful and useful "plain changes" sequence of permutations, plus I never liked determinants[6] to begin with), but it's widely used.

---

[5] John C. Slater (1900–1976), American theoretical physicist who made major contributions to our understanding of atoms, molecules, and solids. Also important as a teacher, textbook author, and administrator.

[6] I am not alone. See Sheldon Axler, "Down with determinants!" *American Mathematical Monthly* **102** (February 1995) 139–154.

6.2.1 *Sample Problem: Try it out*

Find the symmetric and antisymmetric functions generated by the seed functions

$$f(x_A, x_B) = x_A x_B^2 \qquad g(x_A, x_B, x_C) = x_A x_C^2 + 2x_B x_C.$$

What happens if the resulting functions are multiplied by 5, or by -1, or by i?

Solution: For the two-variable function $f(x_A, x_B)$,

$$s(x_A, x_B) = f(x_A, x_B) + f(x_B, x_A) = x_A x_B^2 + x_B x_A^2 = x_A x_B^2 + x_A^2 x_B,$$

$$a(x_A, x_B) = f(x_A, x_B) - f(x_B, x_A) = x_A x_B^2 - x_B x_A^2 = x_A x_B^2 - x_A^2 x_B.$$

And sure enough, just to try some particular cases, $s(5,2) = s(2,5) = 70$, $a(5,2) = -30$, $a(2,5) = +30$, $s(3,3) = 54$, $a(3,3) = 0$. If you multiplied $s(x_A, x_B)$ by 5, or by -1, or by i, or by any number, you would still have a function symmetric under coordinate swaps. Similarly for $a(x_A, x_B)$. Note particularly the multiplication by -1: in definition (6.6) we could have swapped the $+$ and $-$ signs.

For the three-variable function $g(x_A, x_B, x_C)$, equation (6.7) becomes

$$\begin{aligned}
s(x_A, x_B, x_C) = \; & g(x_A, x_B, x_C) \\
& + g(x_A, x_C, x_B) \\
& + g(x_C, x_A, x_B) \\
& + g(x_C, x_B, x_A) \\
& + g(x_B, x_C, x_A) \\
& + g(x_B, x_A, x_C) \\
= \; & x_A x_C^2 + 2x_B x_C \\
& + x_A x_B^2 + 2x_C x_B \\
& + x_C x_B^2 + 2x_A x_B \\
& + x_C x_A^2 + 2x_B x_A \\
& + x_B x_A^2 + 2x_C x_A \\
& + x_B x_C^2 + 2x_A x_C \\
= \; & x_A^2(x_B + x_C) + x_B^2(x_A + x_C) + x_C^2(x_A + x_B) \\
& + 4x_A x_B + 4x_A x_C + 4x_B x_C,
\end{aligned}$$

while equation (6.8) becomes

$$\begin{aligned}
a(x_A, x_B, x_C) = {} & g(x_A, x_B, x_C) \\
& -g(x_A, x_C, x_B) \\
& +g(x_C, x_A, x_B) \\
& -g(x_C, x_B, x_A) \\
& +g(x_B, x_C, x_A) \\
& -g(x_B, x_A, x_C) \\
= {} & x_A x_C^2 + 2x_B x_C \\
& -x_A x_B^2 - 2x_C x_B \\
& +x_C x_B^2 + 2x_A x_B \\
& -x_C x_A^2 - 2x_B x_A \\
& +x_B x_A^2 + 2x_C x_A \\
& -x_B x_C^2 - 2x_A x_C \\
= {} & x_A^2(x_B - x_C) + x_B^2(-x_A + x_C) + x_C^2(x_A - x_B).
\end{aligned}$$

Trying out some particular cases: $s(1,2,3) = s(2,1,3) = s(2,3,1) = 92$, $a(1,2,3) = -a(2,1,3) = a(2,3,1) = -2$, $s(1,1,3) = s(1,3,1) = 54$, $a(1,1,3) = a(1,3,1) = 0$. As in the two-variable case, if we multiply either of these functions by a constant the symmetry or antisymmetry will be unaffected. In the definition (6.8) for the antisymmetrization process we could have swapped the $+$ and $-$ signs.

Problems

6.1 Product functions

Can a product function $F(x_A)G(x_B)$ ever be symmetric under swapping? Antisymmetric?

6.2 Special case

Implement the symmetrization and antisymmetrization procedures (6.7) and (6.8) for the garden-variety function $f(x_A, x_B, x_C) = x_A x_B^2 x_C^3$. Evaluate the resulting functions $s(x_A, x_B, x_C)$ and $a(x_A, x_B, x_C)$ first at $x_A = 1$, $x_B = 2$, and $x_C = 3$, then at $x_A = 3$, $x_B = 2$, and $x_C = 1$. ⟦*Results:* $s(1, 2, 3) = s(3, 2, 1) = 288$, $a(1, 2, 3) = 12$, $a(3, 2, 1) = -12$.⟧

6.3 Antisymmetrizing the symmetric

a. There is one function that is both completely symmetric and completely antisymmetric. What is it?

b. Suppose the seed function is symmetric under a swap of the first two coordinates

$$f(\mathsf{x}_A, \mathsf{x}_B, \mathsf{x}_C) = f(\mathsf{x}_B, \mathsf{x}_A, \mathsf{x}_C)$$

and the antisymmetrization process (6.8) is executed. What is the result?

c. Repeat part (b) for a seed function symmetric under a swap of the last two coordinates.

d. Repeat part (b) for a seed function symmetric under a swap of the first and third coordinates.

e. Suppose the seed function is a product as in equation (6.9), and two of the functions happen to be equal. What is the result of the antisymmetrization process?

6.3 Consequences of the Pauli principle

Does the requirement of symmetry or antisymmetry under coordinate swapping have any consequences? Here's an immediate one for fermions: Take both $\mathsf{x}_A = \mathsf{X}$ and $\mathsf{x}_B = \mathsf{X}$. Now when these coordinates are swapped, you get back to where you started:

$$\psi(\mathsf{X},\mathsf{X}) = -\psi(\mathsf{X},\mathsf{X}) \quad \text{so} \quad \psi(\mathsf{X},\mathsf{X}) = 0. \tag{6.13}$$

Thus, the probability density for two identical fermions to have all the same coordinates is zero.

And here's a consequence for both bosons and fermions. Think about space only, no spin. The (unnormalized) seed function

$$f(x_A, x_B) = e^{-[(x_A-0.5\sigma)^2+(x_B+0.3\sigma)^2]/2\sigma^2}$$

has a maximum when $x_A = 0.5\sigma$ and when $x_B = -0.3\sigma$. This shows up as one hump in the two-variable plots below (drawn taking $\sigma = 1$), which show the normalized probability density proportional to $|f(x_A, x_B)|^2$.

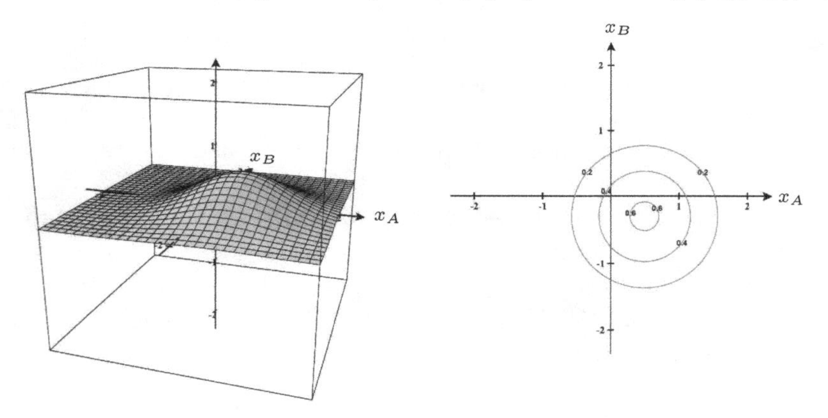

Depending on your background and preferences, you might find it easier to read either the surface plot on the left or the contour plot on the right: both depict the same two-variable function. (And both were drawn using Paul Seeburger's applet CalcPlot3D.)

But what of the symmetric and antisymmetric combinations generated from this seed? Here are surface plots of the normalized probability densities associated with the symmetric (left) and antisymmetric (right) combinations:

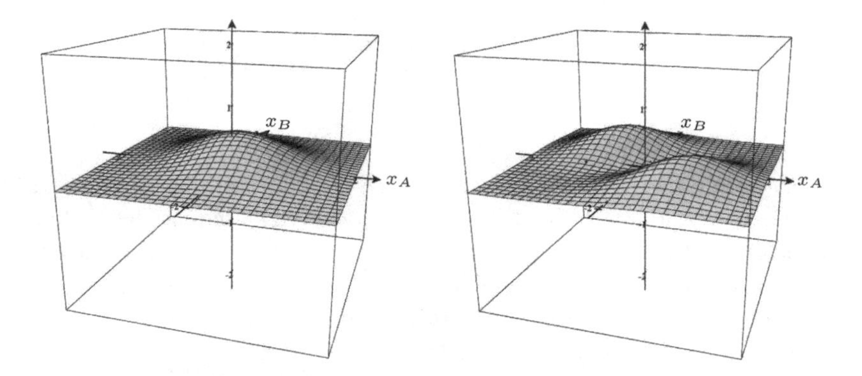

And here are the corresponding contour plots:

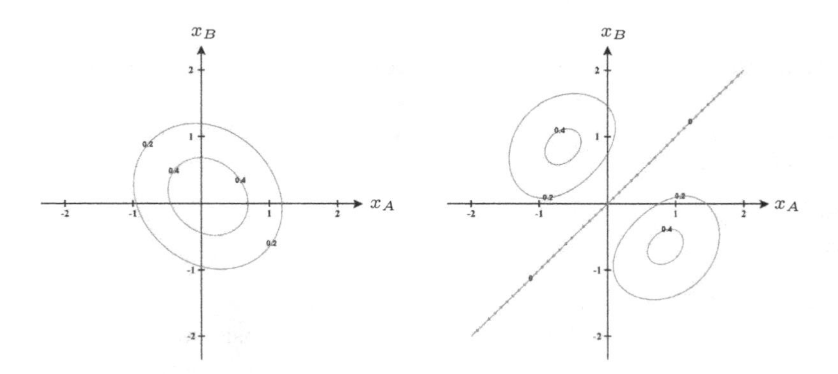

The seed function has no special properties on the $x_A = x_B$ diagonal axis. But, as required by equation (6.13), the antisymmetric combination vanishes there. And the symmetric combination is high there!

The "vanishing on diagonal requirement" and this particular example are but two facets of the more general rule of thumb that:

> In a symmetric spatial wavefunction, the particles tend to huddle together.
> In an antisymmetric spatial wavefunction, the particles tend to spread apart.

This rule is *not* a theorem and you can find counterexamples,[7] but such exceptions are rare.

In everyday experience, when two people tend to huddle together or spread apart, it's for emotional reasons. In everyday experience, when two particles tend to huddle together or spread apart, it's because they're attracted to or repelled from each other through a force. This quantal case is vastly different. The huddling or spreading is of course not caused by emotions and it's also not caused by a force — it occurs for identical particles even when they don't interact. The cause is instead the symmetry/antisymmetry requirement: not a force like a hammer blow, but a piece of mathematics!

Therefore it's difficult to come up with terms for the behavior of identical particles that don't suggest either emotions or forces ascribed to particles: congregate, avoid; gregarious, loner; attract, repel; flock, scatter. "Huddle together" and "spread apart" are the best terms I've been able to devise, but you might be able to find better ones.

Problem

6.4 **Symmetric and antisymmetric combinations**
Two identical particles ambivate in a one-dimensional infinite square well. Take as a seed function the product of energy eigenstates $\eta_2(x_A)\eta_3(x_B)$. Use your favorite graphics package to plot the probability densities associated with the symmetric and antisymmetric combinations generated from this seed. Does the "huddle together/spread apart" rule hold?

[7] See D.F. Styer, "On the separation of identical particles in quantum mechanics" *European Journal of Physics* **41** (14 October 2020) 065402.

6.4 Consequences of the Pauli principle for product states

A commonly encountered special case comes when the many-particle seed function is a product of one-particle functions — we glanced at this special case in equation (6.9). What happens if two of these one-particle functions are the same? Nothing special happens for the symmetrization case. But the answer for antisymmetrization is cute. It pops out of equation (6.11): If $f_1(\mathsf{x}) = f_2(\mathsf{x})$, then the last line cancels the first line, the second cancels the fifth, and the fourth cancels the third. The antisymmetric combination vanishes everywhere!

Unlike the "huddle together/spread apart" rule of thumb, this result *is* a theorem: the antisymmetric combination vanishes if any two of the one-particle functions are the same. It is a partner to the $\mathsf{x}_A = \mathsf{x}_B$ theorem of equation (6.13): just as the two particles can't have the same coordinates, so their wavefunction can't be built from the same one-particle functions.

6.5 Energy states for two identical, noninteracting particles

A **single particle** ambivates subject to some potential energy function. There are M energy eigenstates (where usually $M = \infty$)

$$\eta_1(\mathsf{x}), \eta_2(\mathsf{x}), \eta_3(\mathsf{x}), \ldots, \eta_M(\mathsf{x}). \tag{6.14}$$

Now **two non-identical particles** ambivate subject to the same potential energy. They have the same mass, and do not interact with each other. You can see what the energy eigenstates are: state $\eta_3(\mathsf{x}_A)\eta_8(\mathsf{x}_B)$, for example, has energy $E_3 + E_8$. There's necessarily a degeneracy, as defined on page 150, because the different state $\eta_8(\mathsf{x}_A)\eta_3(\mathsf{x}_B)$ has the same energy. The basis of energy eigenstates has M^2 elements, and they are normalized. Any state can be represented as a linear combination of these elements. I could go on, but the picture is clear: the fact that there are two particles rather than one is unimportant; this basis of energy eigenstates has all the properties you expect of an energy eigenbasis.

This basis consists of product states, but of course that's just a coincidence. You could replace the two basis states

$$\eta_3(\mathsf{x}_A)\eta_8(\mathsf{x}_B) \quad \text{and} \quad \eta_8(\mathsf{x}_A)\eta_3(\mathsf{x}_B)$$

with, for example [using equation (4.39) with $\cos\theta = \frac{4}{5}$],

$$+\tfrac{4}{5}\eta_3(\mathsf{x}_A)\eta_8(\mathsf{x}_B) + \tfrac{3}{5}\eta_8(\mathsf{x}_A)\eta_3(\mathsf{x}_B) \quad \text{and} \quad -\tfrac{3}{5}\eta_3(\mathsf{x}_A)\eta_8(\mathsf{x}_B) + \tfrac{4}{5}\eta_8(\mathsf{x}_A)\eta_3(\mathsf{x}_B).$$

Now **two identical, noninteracting particles** ambivate subject to the same potential energy. In this case we don't want a basis from which we can build *any* wavefunction: we want a basis from which we can build any symmetric wavefunction, or a basis from which we can build any antisymmetric wavefunction.

The basis for antisymmetric wavefunctions has elements like

$$\frac{1}{\sqrt{2}}\left[-\eta_3(\mathsf{x}_A)\eta_8(\mathsf{x}_B)+\eta_8(\mathsf{x}_A)\eta_3(\mathsf{x}_B)\right]. \tag{6.15}$$

The basis for symmetric wavefunctions has elements like

$$\frac{1}{\sqrt{2}}\left[+\eta_3(\mathsf{x}_A)\eta_8(\mathsf{x}_B)+\eta_8(\mathsf{x}_A)\eta_3(\mathsf{x}_B)\right] \tag{6.16}$$

plus elements like

$$\eta_8(\mathsf{x}_A)\eta_8(\mathsf{x}_B). \tag{6.17}$$

You should convince yourself that there are $\frac{1}{2}M(M-1)$ elements in the antisymmetric basis and $\frac{1}{2}M(M+1)$ elements in the symmetric basis. Notice that non-product[8] states are a strict necessity in these bases.

The basis for antisymmetric wavefunctions united with the basis for symmetric wavefunctions produces a basis for any wavefunction. This is a peculiarity of the two-particle case, and reflects the fact that any two-variable function is the sum of a completely symmetric function and a completely antisymmetric function. It is *not* true that any three-variable function is the sum of a completely symmetric and a completely antisymmetric function. For three noninteracting particles, the general basis has M^3 elements, the basis for antisymmetric wavefunctions has $\frac{1}{6}M(M-1)(M-2)$ elements, and the basis for symmetric wavefunctions has $\frac{1}{6}M(M+1)(M+2)$ elements. If you enjoy mathematical puzzles, you will enjoy proving these statements for yourself. But we won't need them for this book.

Before proceeding, I introduce some terminology. The phrase "one-particle states multiplied together then permuted through the symmetrization/antisymmetrization machinery of equations (6.10) and (6.11) to build a many-particle state" is a real mouthful. A "many-particle state" like (6.15) or (6.16) or (6.17) is called just a "state", while a building block

[8]Because the states (6.15 and (6.16) are clearly not in the form of a product $\psi_A(\mathsf{x}_A)\psi_B(\mathsf{x}_B)$, the definition of entangled state on page 129 suggests that these states should be called entangled. However the correct definition of entanglement for identical particles remains unsettled, so I use the term "non-product state" instead.

one-particle state like $\eta_3(x)$ or $\eta_8(x)$ is called a "level".[9] This terminology relieves us of the need say "one-particle" or "many-particle" or "antisymmetrization machinery" over and over again.

The basis for bosons is bigger than the basis for fermions because you can combine levels 1 and 7 to build either a boson state or a fermion state, but you can combine levels 7 and 7 to build a boson state but not a fermion state. As detailed in the previous section, "Consequences of the Pauli principle for product states", the levels combined to make a fermion state must all be different.

Problem

6.5 **Building three-particle basis states**
Suppose you had three particles and three "building block" levels (say the orthonormal levels $\eta_1(\mathsf{x})$, $\eta_3(\mathsf{x})$, and $\eta_7(\mathsf{x})$). Construct normalized three-particle basis states for the case of

a. three non-identical particles
b. three identical bosons
c. three identical fermions

How many states are there in each basis? Repeat for three particles with four one-particle levels, but in this case simply count and don't write down all the three-particle states.

6.6 Spin plus space, two electrons

Electrons are spin-half fermions. Two of them ambivate subject to the same potential. Energy doesn't depend on spin. Pretend the two electrons don't interact. (Perhaps a better name for this section would be "Spin plus space, two noninteracting spin-$\frac{1}{2}$ fermions", but yikes, how long do you want this section's title to be? Should I add "non-relativistic" and "ignoring collisions" and "ignoring radiation"?)

The spatial energy levels for one electron are $\eta_n(\vec{x})$ for $n = 1, 2, \ldots, M/2$. Thus the full (spin plus space) energy levels for one electron are the M levels

[9] Some people, particularly chemists referring to atomic systems, use the term "orbital" rather than "level". This term unfortunately suggests a circular Bohr orbit. An electron with an energy does *not* execute a circular Bohr orbit at constant speed. Instead it ambivates without position or velocity.

$\eta_n(\vec{x})\chi_+$ and $\eta_n(\vec{x})\chi_-$. Now the question: What are the energy eigenstates for the two noninteracting electrons?

Well, what two-particle states can we build from the one-particle spatial levels with, say, $n = 1$ and $n = 3$? (Once you see how to do it for $n = 1$ and $n = 3$, you can readily generalize to any two values of n.) These correspond to four levels:

$$\eta_1(\vec{x})\chi_+, \tag{6.18}$$

$$\eta_1(\vec{x})\chi_-, \tag{6.19}$$

$$\eta_3(\vec{x})\chi_+, \tag{6.20}$$

$$\eta_3(\vec{x})\chi_-. \tag{6.21}$$

What states mixing $n = 1$ with $n = 3$ can be built from these four levels?

The antisymmetric combination of (6.18) with itself vanishes. The antisymmetric combination of (6.18) with (6.19) is a combination of $n = 1$ with $n = 1$, not of $n = 1$ with $n = 3$. The (unnormalzed) antisymmetric combination of (6.18) with (6.20) is

$$\eta_1(\vec{x}_A)\chi_+(A)\eta_3(\vec{x}_B)\chi_+(B) - \eta_3(\vec{x}_A)\chi_+(A)\eta_1(\vec{x}_B)\chi_+(B). \tag{6.22}$$

The antisymmetric combination of (6.18) with (6.21) is

$$\eta_1(\vec{x}_A)\chi_+(A)\eta_3(\vec{x}_B)\chi_-(B) - \eta_3(\vec{x}_A)\chi_-(A)\eta_1(\vec{x}_B)\chi_+(B). \tag{6.23}$$

The antisymmetric combination of (6.19) with (6.20) is

$$\eta_1(\vec{x}_A)\chi_-(A)\eta_3(\vec{x}_B)\chi_+(B) - \eta_3(\vec{x}_A)\chi_+(A)\eta_1(\vec{x}_B)\chi_-(B). \tag{6.24}$$

The antisymmetric combination of (6.19) with (6.21) is

$$\eta_1(\vec{x}_A)\chi_-(A)\eta_3(\vec{x}_B)\chi_-(B) - \eta_3(\vec{x}_A)\chi_-(A)\eta_1(\vec{x}_B)\chi_-(B). \tag{6.25}$$

Finally, the antisymmetric combination of (6.20) with (6.21) is a combination of $n = 3$ with $n = 3$, not of $n = 1$ with $n = 3$.

All four of these states are energy eigenstates with energy $E_1 + E_3$. State (6.22) factorizes into a convenient space-times-spin form:

$$\begin{aligned} &\eta_1(\vec{x}_A)\chi_+(A)\eta_3(\vec{x}_B)\chi_+(B) - \eta_3(\vec{x}_A)\chi_+(A)\eta_1(\vec{x}_B)\chi_+(B) \\ &= \Big[\eta_1(\vec{x}_A)\eta_3(\vec{x}_B) - \eta_3(\vec{x}_A)\eta_1(\vec{x}_B)\Big]\,\chi_+(A)\chi_+(B). \end{aligned} \tag{6.26}$$

The space part of the wavefunction is antisymmetric under coordinate swap. The spin part is symmetric. Thus the total wavefunction is antisymmetric.

Before proceeding I confess that I'm sick and tired of writing all these ηs and χs and As and Bs that convey no information. I *always* write the η in front of the χ. I *always* write the As in front of the Bs. You'll *never* confuse an η with a χ, because the ηs are labeled 1, 3 while the χs are labeled $+$, $-$. Dirac introduced a notation (see page 90) that takes all this for granted, so that neither you nor I have to write the same thing out over and over again. This notation usually replaces $+$ with $\uparrow$ and $-$ with $\downarrow$ (see page 112). In this notation, equation (6.26) is written

$$|1\uparrow,3\uparrow\rangle - |3\uparrow,1\uparrow\rangle = \Big[|1,3\rangle - |3,1\rangle\Big]\,|\uparrow\uparrow\rangle. \tag{6.27}$$

In this new notation the states (6.22) through (6.25) are written

$$\Big[|1,3\rangle - |3,1\rangle\Big]\,|\uparrow\uparrow\rangle \tag{6.28}$$

$$|1\uparrow,3\downarrow\rangle - |3\downarrow,1\uparrow\rangle \tag{6.29}$$

$$|1\downarrow,3\uparrow\rangle - |3\uparrow,1\downarrow\rangle \tag{6.30}$$

$$\Big[|1,3\rangle - |3,1\rangle\Big]\,|\downarrow\downarrow\rangle. \tag{6.31}$$

Well, this is cute. Two of the four states have this convenient space-times-spin form... and furthermore these two have the same spatial wavefunction! Two other states, however, don't have this convenient form.

One thing to do about this is nothing. There's no requirement that states have a space-times-spin form. But in this two-electron case there's a slick trick that enables us to put the states into space-times-spin form.

Because all four states (6.28) through (6.31) have the same energy, namely $E_1 + E_3$, I can make linear combinations of the states to form other equally good energy states. Can I make a combination of states (6.29) and (6.30) that *does* factorize into space times spin? Nothing ventured, nothing gained. Let's try it:

$$\alpha\Big[|1\uparrow,3\downarrow\rangle - |3\downarrow,1\uparrow\rangle\Big] + \beta\Big[|1\downarrow,3\uparrow\rangle - |3\uparrow,1\downarrow\rangle\Big]$$
$$= |1,3\rangle\Big[\alpha|\uparrow\downarrow\rangle + \beta|\downarrow\uparrow\rangle\Big] - |3,1\rangle\Big[\alpha|\downarrow\uparrow\rangle + \beta|\uparrow\downarrow\rangle\Big].$$

This will factorize only if the left term in square brackets is proportional to the right term in square brackets:

$$\Big[\alpha|\uparrow\downarrow\rangle + \beta|\downarrow\uparrow\rangle\Big] = c\Big[\beta|\uparrow\downarrow\rangle + \alpha|\downarrow\uparrow\rangle\Big],$$

that is only if

$$\alpha = c\beta \qquad \text{and} \qquad \beta = c\alpha.$$

Combining these two equations results in $c = \pm 1$. If $c = +1$ then the combination results in the state

$$\Big[|1,3\rangle - |3,1\rangle\Big]\,\alpha\,\Big[|\uparrow\downarrow\rangle + |\downarrow\uparrow\rangle\Big], \tag{6.32}$$

whereas when $c = -1$ the result is

$$\Big[|1,3\rangle + |3,1\rangle\Big]\,\alpha\,\Big[|\uparrow\downarrow\rangle - |\downarrow\uparrow\rangle\Big]. \tag{6.33}$$

Putting all this together and, for the sake of good form, insuring normalized states, we find that the two-electron energy states in equations (6.28) through (6.31) can be recast as

$$\Big[\tfrac{1}{\sqrt{2}}(|1,3\rangle - |3,1\rangle)\Big]\,|\uparrow\uparrow\rangle \tag{6.34}$$

$$\Big[\tfrac{1}{\sqrt{2}}(|1,3\rangle - |3,1\rangle)\Big]\,\Big[\tfrac{1}{\sqrt{2}}(|\uparrow\downarrow\rangle + |\downarrow\uparrow\rangle)\Big] \tag{6.35}$$

$$\Big[\tfrac{1}{\sqrt{2}}(|1,3\rangle - |3,1\rangle)\Big]\,|\downarrow\downarrow\rangle \tag{6.36}$$

$$\Big[\tfrac{1}{\sqrt{2}}(|1,3\rangle + |3,1\rangle)\Big]\,\Big[\tfrac{1}{\sqrt{2}}(|\uparrow\downarrow\rangle - |\downarrow\uparrow\rangle)\Big]. \tag{6.37}$$

The first three of these states have spatial wavefunctions antisymmetric under coordinate swaps and spin wavefunctions symmetric under coordinate swaps — these are called "ortho states" or "a triplet". The last one has a symmetric spatial wavefunction and an antisymmetric spin wavefunction — these are called "para states" or "a singlet". Our discussion in section 6.3, "Consequences of the Pauli principle", demonstrates that in ortho states, the two electrons tend to spread apart in space; in para states, they tend to huddle together.

I write out the singlet spin state

$$\tfrac{1}{\sqrt{2}}\left[|\uparrow\downarrow\rangle - |\downarrow\uparrow\rangle\right] \tag{6.38}$$

using the verbose terminology

$$\tfrac{1}{\sqrt{2}}\left[\chi_+(A)\chi_-(B) - \chi_-(A)\chi_+(B)\right] \tag{6.39}$$

to make it absolutely clear that coordinate A is associated with both spin $+$ and spin $-$, as is coordinate B. It is impossible to say that "one electron has spin up and the other has spin down".

This abstract machinery might seem purely formal, but in fact it has tangible experimental consequences. In the sample problem below, the machinery suggests that the ground state of the hydrogen atom is two-fold degenerate, while the ground state of the helium atom is non-degenerate. And this prediction is borne out by experiment!

6.6.1 *Sample Problem: Ground state degeneracy for one and two electrons*

A certain potential energy function has two spatial energy eigenstates: $\eta_1(\vec{x})$ with energy E_1 and $\eta_2(\vec{x})$ with a higher energy E_2. These energies are independent of spin.

a. A single electron (spin-$\frac{1}{2}$) ambivates in this potential. Write out the four energy eigenstates and the energy eigenvalue associated with each. What is the ground state degeneracy?
b. Two non-interacting electrons ambivate in this same potential. Write out the six energy eigenstates and the energy eigenvalue associated with each. What is the ground state degeneracy?

Solution: **(a)** For the single electron:

energy eigenstate	energy eigenvalue
$\eta_1(\vec{x})\chi_+$	E_1
$\eta_1(\vec{x})\chi_-$	E_1
$\eta_2(\vec{x})\chi_+$	E_2
$\eta_2(\vec{x})\chi_-$	E_2

The first two states listed are both ground states, so the ground state is two-fold degenerate.

(b) For the two electrons, we build states from levels just as we did in this section. The first line below is the antisymmetrized combination of $\eta_1(\vec{x})\chi_+$ with $\eta_1(\vec{x})\chi_-$. This state has energy $2E_1$. The next four lines are built up exactly as equations (6.34) through (6.37) were. Each of these four states has energy E_1+E_2. The last line is the antisymmetrized combination of $\eta_2(\vec{x})\chi_+$ with $\eta_2(\vec{x})\chi_-$. This state has energy $2E_2$.

$$\eta_1(\vec{x}_A)\eta_1(\vec{x}_B)\tfrac{1}{\sqrt{2}}\left[\chi_+(A)\chi_-(B) - \chi_-(A)\chi_+(B)\right]$$
$$\tfrac{1}{\sqrt{2}}\left[\eta_1(\vec{x}_A)\eta_2(\vec{x}_B) - \eta_2(\vec{x}_A)\eta_1(\vec{x}_B)\right]\left[\chi_+(A)\chi_+(B)\right]$$
$$\tfrac{1}{\sqrt{2}}\left[\eta_1(\vec{x}_A)\eta_2(\vec{x}_B) - \eta_2(\vec{x}_A)\eta_1(\vec{x}_B)\right]\tfrac{1}{\sqrt{2}}\left[\chi_+(A)\chi_-(B) + \chi_-(A)\chi_+(B)\right]$$
$$\tfrac{1}{\sqrt{2}}\left[\eta_1(\vec{x}_A)\eta_2(\vec{x}_B) - \eta_2(\vec{x}_A)\eta_1(\vec{x}_B)\right]\left[\chi_-(A)\chi_-(B)\right]$$
$$\tfrac{1}{\sqrt{2}}\left[\eta_1(\vec{x}_A)\eta_2(\vec{x}_B) + \eta_2(\vec{x}_A)\eta_1(\vec{x}_B)\right]\tfrac{1}{\sqrt{2}}\left[\chi_+(A)\chi_-(B) - \chi_-(A)\chi_+(B)\right]$$
$$\eta_2(\vec{x}_A)\eta_2(\vec{x}_B)\tfrac{1}{\sqrt{2}}\left[\chi_+(A)\chi_-(B) - \chi_-(A)\chi_+(B)\right].$$

The ground state of the two-electron system is the first state listed: it is non-degenerate.

Problems

6.6 **Combining a spatial one-particle level with itself**
What two-particle states can we build from the one-particle spatial level with $n = 3$? How many of the resulting states are ortho, how many para?

6.7 **Change of basis through abstract rotation**
Show that, in retrospect, the process of building states (6.35) and (6.37) from states (6.29) and (6.30) is nothing but a "45° rotation" in the style of equation (4.39).

6.8 **Normalization of singlet spin state**
Justify the normalization constant $\frac{1}{\sqrt{2}}$ that enters in moving from equation (6.33) to equation (6.37). Compare this singlet spin state to the entangled state (3.37). (Indeed, one way to produce an entangled pair of electrons is to start in a singlet state and then draw the two electrons apart.)

6.9 **Ortho and para accounting**
Show that in our case with $M/2$ spatial energy levels, the two-electron energy basis has $\frac{1}{2}M(M-1)$ elements, of which

$$\tfrac{3}{2}(M/2)[(M/2)-1] \text{ are ortho}$$

(antisymmetric in space and symmetric in spin) and

$$\tfrac{1}{2}(M/2)[(M/2)+1] \text{ are para}$$

(symmetric in space and antisymmetric in spin).

6.10 Intersystem crossing

A one-electron system has a ground level $\eta_g(\vec{x})$ and an excited level $\eta_e(\vec{x})$, for a total of four basis levels:

$$\eta_g(\vec{x})\chi_+, \quad \eta_g(\vec{x})\chi_-, \quad \eta_e(\vec{x})\chi_+, \quad \eta_e(\vec{x})\chi_-.$$

A basis for two-electron states is then the six states:

$$\eta_g(\vec{x}_A)\eta_g(\vec{x}_B)\frac{1}{\sqrt{2}}\left[\chi_+(A)\chi_-(B) - \chi_-(A)\chi_+(B)\right]$$
$$\frac{1}{\sqrt{2}}\left[\eta_g(\vec{x}_A)\eta_e(\vec{x}_B) - \eta_e(\vec{x}_A)\eta_g(\vec{x}_B)\right]\left[\chi_+(A)\chi_+(B)\right]$$
$$\frac{1}{\sqrt{2}}\left[\eta_g(\vec{x}_A)\eta_e(\vec{x}_B) - \eta_e(\vec{x}_A)\eta_g(\vec{x}_B)\right]\frac{1}{\sqrt{2}}\left[\chi_+(A)\chi_-(B) + \chi_-(A)\chi_+(B)\right]$$
$$\frac{1}{\sqrt{2}}\left[\eta_g(\vec{x}_A)\eta_e(\vec{x}_B) - \eta_e(\vec{x}_A)\eta_g(\vec{x}_B)\right]\left[\chi_-(A)\chi_-(B)\right]$$
$$\frac{1}{\sqrt{2}}\left[\eta_g(\vec{x}_A)\eta_e(\vec{x}_B) + \eta_e(\vec{x}_A)\eta_g(\vec{x}_B)\right]\frac{1}{\sqrt{2}}\left[\chi_+(A)\chi_-(B) - \chi_-(A)\chi_+(B)\right]$$
$$\eta_e(\vec{x}_A)\eta_e(\vec{x}_B)\frac{1}{\sqrt{2}}\left[\chi_+(A)\chi_-(B) - \chi_-(A)\chi_+(B)\right].$$

A transition from the second state listed above to the first is called an "intersystem crossing". One sometimes reads, in association with the diagram below, that in an intersystem crossing "the spin of the excited electron is reversed". In five paragraphs or fewer, explain why this phrase is inaccurate, perhaps even grotesque, and suggest a replacement.

6.7 Spin plus space, three electrons, ground state

Three electrons are in the situation described in the first paragraph of section 6.6 (energy independent of spin, electrons don't interact). The full listing of energy eigenstates has been done, but it's an accounting nightmare, so I ask a simpler question: What is the ground state?

Call the one-particle spatial energy levels $\eta_1(\vec{x})$, $\eta_2(\vec{x})$, $\eta_3(\vec{x})$, ... The ground state will be the antisymmetrized combination of the three levels

$$\eta_1(\vec{x}_A)\chi_+(A) \qquad \eta_1(\vec{x}_B)\chi_-(B) \qquad \eta_2(\vec{x}_C)\chi_+(C)$$

or the antisymmetrized combination of the three levels

$$\eta_1(\vec{x}_A)\chi_+(A) \qquad \eta_1(\vec{x}_B)\chi_-(B) \qquad \eta_2(\vec{x}_C)\chi_-(C).$$

The two states so generated are degenerate:[10] both have energy $2E_1 + E_2$.

Write out the first state in detail. It is

$$\begin{aligned}\tfrac{1}{\sqrt{6}}[\ & \eta_1(\vec{x}_A)\chi_+(A)\,\eta_1(\vec{x}_B)\chi_-(B)\,\eta_2(\vec{x}_C)\chi_+(C)\\ & -\eta_1(\vec{x}_A)\chi_+(A)\,\eta_2(\vec{x}_B)\chi_+(B)\,\eta_1(\vec{x}_C)\chi_-(C)\\ & +\eta_2(\vec{x}_A)\chi_+(A)\,\eta_1(\vec{x}_B)\chi_+(B)\,\eta_1(\vec{x}_C)\chi_-(C)\\ & -\eta_2(\vec{x}_A)\chi_+(A)\,\eta_1(\vec{x}_B)\chi_-(B)\,\eta_1(\vec{x}_C)\chi_+(C)\\ & +\eta_1(\vec{x}_A)\chi_-(A)\,\eta_2(\vec{x}_B)\chi_+(B)\,\eta_1(\vec{x}_C)\chi_+(C)\\ & -\eta_1(\vec{x}_A)\chi_-(A)\,\eta_1(\vec{x}_B)\chi_+(B)\,\eta_2(\vec{x}_C)\chi_+(C)\ ].\end{aligned} \tag{6.40}$$

This morass is another good argument for the abbreviated Dirac notation introduced on page 221. I'm not concerned with normalization for the moment, so I'll write this first state as

$$\begin{aligned}&|1\uparrow, 1\downarrow, 2\uparrow\rangle\\ -&|1\uparrow, 2\uparrow, 1\downarrow\rangle\\ +&|2\uparrow, 1\uparrow, 1\downarrow\rangle\\ -&|2\uparrow, 1\downarrow, 1\uparrow\rangle\\ +&|1\downarrow, 2\uparrow, 1\uparrow\rangle\\ -&|1\downarrow, 1\uparrow, 2\uparrow\rangle\end{aligned} \tag{6.41}$$

and the second one (with $2\downarrow$ replacing $2\uparrow$) as

$$\begin{aligned}&|1\uparrow, 1\downarrow, 2\downarrow\rangle\\ -&|1\uparrow, 2\downarrow, 1\downarrow\rangle\\ +&|2\downarrow, 1\uparrow, 1\downarrow\rangle\\ -&|2\downarrow, 1\downarrow, 1\uparrow\rangle\\ +&|1\downarrow, 2\downarrow, 1\uparrow\rangle\\ -&|1\downarrow, 1\uparrow, 2\downarrow\rangle.\end{aligned} \tag{6.42}$$

Both of these states are antisymmetric, but neither factorizes into a neat "space part times spin part". If, following the approach used with two electrons, you attempt to find a linear combination of these two that does so factorize, you will fail: see problem 6.11. The ground state wavefunction cannot be made to factor into a space part times a spin part.

[10]See the definition on page 150 and problem 4.11 on page 151.

Problems

6.11 **A doomed attempt** (essential problem)
Any linear combination of state (6.41) with state (6.42) has the form

$$\begin{aligned}
&|1,1,2\rangle \left[\alpha|\uparrow\downarrow\uparrow\rangle + \beta|\uparrow\downarrow\downarrow\rangle\right] \\
-&|1,2,1\rangle \left[\alpha|\uparrow\uparrow\downarrow\rangle + \beta|\uparrow\downarrow\downarrow\rangle\right] \\
+&|2,1,1\rangle \left[\alpha|\uparrow\uparrow\downarrow\rangle + \beta|\downarrow\uparrow\downarrow\rangle\right] \\
-&|2,1,1\rangle \left[\alpha|\uparrow\downarrow\uparrow\rangle + \beta|\downarrow\downarrow\uparrow\rangle\right] \\
+&|1,2,1\rangle \left[\alpha|\downarrow\uparrow\uparrow\rangle + \beta|\downarrow\downarrow\uparrow\rangle\right] \\
-&|1,1,2\rangle \left[\alpha|\downarrow\uparrow\uparrow\rangle + \beta|\downarrow\uparrow\downarrow\rangle\right].
\end{aligned} \tag{6.43}$$

Show that this form can never be factorized into a space part times a spin part.

6.12 **Questions** (recommended problem)
Update your list of quantum mechanics questions that you started at problem 1.17 on page 46. Write down new questions and, if you have uncovered answers to any of your old questions, write them down briefly.

Chapter 7

Atoms

During the months following these discussions [in the autumn of 1926] an intensive study of all questions concerning the interpretation of quantum theory in Copenhagen finally led to a complete and, as many physicists believe, satisfactory clarification of the situation. But it was not a solution which one could easily accept. I remember discussions with Bohr which went through many hours till very late at night and ended almost in despair; and when at the end of the discussion I went alone for a walk in the neighboring park I repeated to myself again and again the question: Can nature possibly be as absurd as it seemed to us in these atomic experiments?

— Werner Heisenberg, *Physics and Philosophy*
(Harper, New York, 1958) page 42

All this is fine and good — lovely, in fact. But we have to apply quantum mechanics to experimentally accessible systems, and while things like carbon nanotubes exist, the most readily accessible systems are atoms.

7.1 Central potentials in two dimensions

Before jumping directly to three-dimensional atoms, we test out the mathematics in two dimensions.

In one dimension, the energy eigenproblem is

$$-\frac{\hbar^2}{2M}\frac{d^2\eta(x)}{dx^2} + V(x)\eta(x) = E\eta(x). \tag{7.1}$$

The generalization to two dimensions is straightforward:

$$-\frac{\hbar^2}{2M}\left[\frac{\partial^2\eta(x,y)}{\partial x^2} + \frac{\partial^2\eta(x,y)}{\partial y^2}\right] + V(x,y)\eta(x,y) = E\eta(x,y). \tag{7.2}$$

The part in square brackets is called "the Laplacian of $\eta(x, y)$" and is represented by the symbol "∇^2" as follows

$$\left[\frac{\partial^2 f(x,y)}{\partial x^2} + \frac{\partial^2 f(x,y)}{\partial y^2}\right] \equiv \nabla^2 f(x,y). \tag{7.3}$$

Thus the "mathematical form" of the energy eigenproblem is

$$\nabla^2 \eta(\vec{r}) + \frac{2M}{\hbar^2}[E - V(\vec{r})]\eta(\vec{r}) = 0. \tag{7.4}$$

Suppose $V(x, y)$ is a "central potential" — that is, a function of distance from the origin r only. Then it makes sense to use polar coordinates r and θ rather than Cartesian coordinates x and y. What is the expression for the Laplacian in polar coordinates? This can be uncovered through the chain rule, and it's pretty hard to do. Fortunately, you can look up the answer:

$$\nabla^2 f(\vec{r}) = \left[\frac{1}{r}\frac{\partial}{\partial r}\left(r\frac{\partial f(r,\theta)}{\partial r}\right) + \frac{1}{r^2}\frac{\partial^2 f(r,\theta)}{\partial \theta^2}\right]. \tag{7.5}$$

Thus, the partial differential equation to be solved is

$$\left[\frac{1}{r}\frac{\partial}{\partial r}\left(r\frac{\partial \eta(r,\theta)}{\partial r}\right) + \frac{1}{r^2}\frac{\partial^2 \eta(r,\theta)}{\partial \theta^2}\right] + \frac{2M}{\hbar^2}[E - V(r)]\eta(r,\theta) = 0 \tag{7.6}$$

or

$$\frac{\partial^2 \eta(r,\theta)}{\partial \theta^2} + r\frac{\partial}{\partial r}\left(r\frac{\partial \eta(r,\theta)}{\partial r}\right) + \frac{2M}{\hbar^2}r^2[E - V(r)]\eta(r,\theta) = 0. \tag{7.7}$$

Use the "separation of variables" strategy introduced on page 133 : look for solutions of the product form

$$\eta(r,\theta) = R(r)\Theta(\theta), \tag{7.8}$$

and hope against hope that all the solutions (or at least some of them) will be of this form. Plugging this product form into the PDE gives

$$R(r)\Theta''(\theta) + \Theta(\theta)\left\{r\frac{d}{dr}\left(r\frac{dR(r)}{dr}\right) + \frac{2M}{\hbar^2}r^2[E - V(r)]R(r)\right\} = 0$$
$$\frac{\Theta''(\theta)}{\Theta(\theta)} + \left\{\frac{r}{R(r)}\frac{d}{dr}\left(r\frac{dR(r)}{dr}\right) + \frac{2M}{\hbar^2}r^2[E - V(r)]\right\} = 0. \tag{7.9}$$

Through the usual separation-of-variables argument, we recognize that if a function of θ alone plus a function of r alone sum to zero, where θ and r are independent variables, then both functions must be equal to a constant:

$$\frac{r}{R(r)}\frac{d}{dr}\left(r\frac{dR(r)}{dr}\right) + \frac{2M}{\hbar^2}r^2[E - V(r)] = -\frac{\Theta''(\theta)}{\Theta(\theta)} = \text{const.} \tag{7.10}$$

First, look at the angular part:

$$\Theta''(\theta) = -\text{const}\,\Theta(\theta). \tag{7.11}$$

This is the differential equation for a mass on a spring! We've already examined it at equations (4.17) and (5.2). The two linearly independent solutions are

$$\Theta(\theta) = \sin(\sqrt{\text{const}}\,\theta) \quad \text{or} \quad \Theta(\theta) = \cos(\sqrt{\text{const}}\,\theta). \tag{7.12}$$

Now, the boundary condition for this ODE is just that the function must come back to itself if θ increases by 2π:

$$\Theta(\theta) = \Theta(2\pi + \theta). \tag{7.13}$$

If you think about this for a minute, you'll see that this means $\sqrt{\text{const}}$ must be an integer. You'll also see that negative integers don't give us anything new, so we'll take

$$\sqrt{\text{const}} = \ell \quad \text{where} \quad \ell = 0, 1, 2, \ldots. \tag{7.14}$$

In summary, the solution to the angular problem is

	$\ell = 0$	$\ell = 1$	$\ell = 2$	$\ell = 3$	$\cdots$
$\Theta(\theta)$	1	$\sin\theta$ or $\cos\theta$	$\sin 2\theta$ or $\cos 2\theta$	$\sin 3\theta$ or $\cos 3\theta$	$\cdots$

Now examine the radial part of the problem:

$$\frac{r}{R(r)}\frac{d}{dr}\left(r\frac{dR(r)}{dr}\right) + \frac{2M}{\hbar^2}r^2[E - V(r)] = \text{const} = \ell^2 \tag{7.15}$$

or, after some manipulation,

$$\frac{1}{r}\frac{d}{dr}\left(r\frac{dR(r)}{dr}\right) + \frac{2M}{\hbar^2}\left[E - V(r) - \frac{\hbar^2}{2M}\frac{\ell^2}{r^2}\right]R(r) = 0. \tag{7.16}$$

Compare this differential equation with another one-variable differential equation, namely the one for the energy eigenproblem in one dimension:

$$\frac{d^2\eta(x)}{dx^2} + \frac{2M}{\hbar^2}[E - V(x)]\,\eta(x) = 0. \tag{7.17}$$

The parts to the right are rather similar, but the parts to the left — the derivatives — are rather different. In addition, the one-dimensional energy eigenfunction satisfies the normalization

$$\int_{-\infty}^{\infty} |\eta(x)|^2\,dx = 1, \tag{7.18}$$

whereas the two-dimensional energy eigenfunction satisfies the normalization

$$\begin{aligned}
\int |\eta(x,y)|^2\, dx\, dy &= 1 \\
\int_0^\infty dr \int_0^{2\pi} r\, d\theta\, |R(r)\sin(\ell\theta)|^2 &= 1 \\
\pi \int_0^\infty dr\, r|R(r)|^2 &= 1.
\end{aligned} \tag{7.19}$$

This suggests that the true analog of the one-dimensional $\eta(x)$ is not $R(r)$, but rather

$$u(r) = \sqrt{r}R(r). \tag{7.20}$$

Furthermore,

$$\text{if} \quad u(r) = \sqrt{r}R(r), \quad \text{then} \quad \frac{1}{r}\frac{d}{dr}\left(r\frac{dR(r)}{dr}\right) = \frac{1}{\sqrt{r}}\left(u''(r) + \frac{1}{4}\frac{u(r)}{r^2}\right). \tag{7.21}$$

Using this change of function, the radial equation (7.16) becomes

$$\begin{aligned}
\frac{d^2u(r)}{dr^2} + \frac{1}{4}\frac{u(r)}{r^2} + \frac{2M}{\hbar^2}\left[E - V(r) - \frac{\hbar^2}{2M}\frac{\ell^2}{r^2}\right] u(r) &= 0, \\
\frac{d^2u(r)}{dr^2} + \frac{2M}{\hbar^2}\left[E - V(r) - \frac{\hbar^2(\ell^2 - \frac{1}{4})}{2M}\frac{1}{r^2}\right] u(r) &= 0.
\end{aligned} \tag{7.22}$$

In this form, the radial equation is exactly like a one-dimensional energy eigenproblem, except that where the one-dimensional problem has the function $V(x)$, the radial problem has the function $V(r) + \hbar^2(\ell^2 - \frac{1}{4})/(2Mr^2)$. These two functions play parallel mathematical roles in the two problems. To emphasize these similar roles, we define an "effective potential energy function" for the radial problem, namely

$$V_{\text{eff}}(r) = V(r) + \frac{\hbar^2(\ell^2 - \frac{1}{4})}{2M}\frac{1}{r^2}. \tag{7.23}$$

Don't read too much into the term "effective potential energy". No actual potential energy function will depend upon $\hbar$, still less upon the separation constant ℓ! I'm not saying that $V_{\text{eff}}(r)$ *is* a potential energy function, merely that it plays the mathematical role of one in solving this one-dimensional eigenproblem.

Now that the radial equation (7.22) is in exact correspondence with the one-dimensional equation (7.17), we can solve this eigenproblem using either of the techniques described in chapter 5, "Solving the Energy Eigenproblem". (Or any other technique that works for the one-dimensional problem.) The resulting eigenfunctions and eigenvalues will, of course, depend upon the value of the separation constant ℓ, because the effective potential depends upon ℓ. And as always, for each ℓ there will be many eigenfunctions and eigenvalues, which we will label by index $n = 1, 2, 3, \ldots$ calling them $u_{n,\ell}(r)$ with eigenvalue $E_{n,\ell}$.

So we see how to find an infinite number of solutions to the partial differential eigenproblem (7.7). The question is, did we get all of them? The answer is in fact "yes," although that's not at all obvious. If you want to learn more, you will need to read up on PDEs and Sturm-Liouville theory!

Summary: To solve the two-dimensional energy eigenproblem for a radially symmetric potential energy function $V(r)$, namely

$$-\frac{\hbar^2}{2M}\nabla^2\eta(\vec{r}) + V(r)\eta(\vec{r}) = E\eta(\vec{r}), \tag{7.24}$$

first solve the one-dimensional radial energy eigenproblem

$$-\frac{\hbar^2}{2M}\frac{d^2u(r)}{dr^2} + \left[V(r) + \frac{\hbar^2(\ell^2 - \frac{1}{4})}{2M}\frac{1}{r^2}\right]u(r) = Eu(r) \tag{7.25}$$

for $\ell = 0, 1, 2, \ldots$. For a given ℓ, call the resulting energy eigenfunctions and eigenvalues $u_{n,\ell}(r)$ and $E_{n,\ell}$ for $n = 1, 2, 3, \ldots$. Then the two-dimensional solutions are

$$\text{for } \ell = 0: \quad \eta(r,\theta) = \frac{u_{n,0}(r)}{\sqrt{r}} \quad \text{with energy } E_{n,0} \tag{7.26}$$

and

$$\text{for } \ell = 1, 2, 3, \ldots: \quad \begin{array}{c} \eta(r,\theta) = \dfrac{u_{n,\ell}(r)}{\sqrt{r}}\sin(\ell\theta) \\ \text{and} \\ \eta(r,\theta) = \dfrac{u_{n,\ell}(r)}{\sqrt{r}}\cos(\ell\theta) \end{array} \quad \text{with energy } E_{n,\ell}. \tag{7.27}$$

Remark 1: For $\ell \neq 0$, there are two different eigenfunctions attached to the same eigenvalue, a situation called degeneracy.[1] Degeneracy is not merely an abstraction concocted by air-head theorists. It can be uncovered experimentally through the intensity — although not the wavelength — of

[1] See the definition on page 150 and problem 4.11 on page 151.

spectral lines, through statistical mechanical effects, and through Zeeman splitting when the atom is placed in a magnetic field.

Remark 2: The energy eigenvalues $E_{n,\ell}$ come about from solving the one-variable energy eigenproblem with effective potential

$$V_{\text{eff}}(r) = V(r) + \frac{\hbar^2(\ell^2 - \frac{1}{4})}{2M}\frac{1}{r^2}.$$

Now, it's clear from inspection that for any value of r, $V_{\text{eff}}(r)$ increases with increasing ℓ. It's reasonable then that the energy eigenvalues also increase with increasing ℓ: that the fifth eigenvalue, for example, will always satisfy $E_{5,0} < E_{5,1} < E_{5,2}$ and so forth. This guess is in fact correct, and it can be proven mathematically, but it's so reasonable that I won't interrupt this story to prove it.

Remark 3: The conventional choice of zero level for a potential energy function is to set $V(r) = 0$ as $r \to \infty$. Hence all of the bound-state energy eigenvalues are expected to be negative.

In summary, the energy eigenvalues for some generic two-dimensional radially symmetric potential will look sort of like this (showing only the four lowest energy eigenvalues for each value of ℓ):

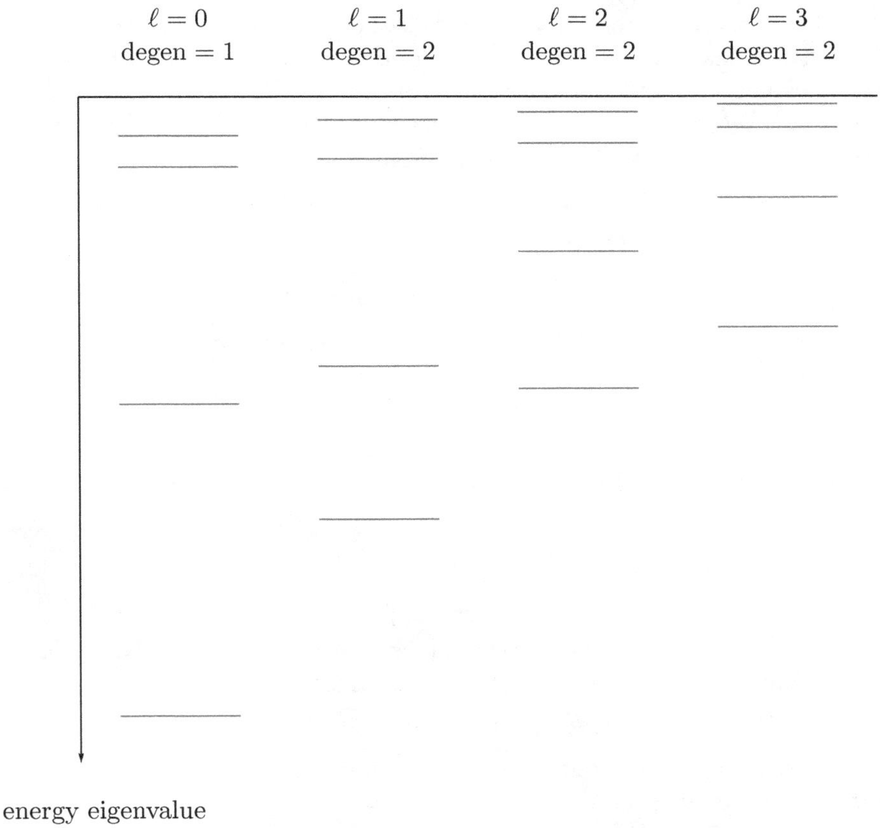

$\ell = 0$ degen = 1 $\quad$ $\ell = 1$ degen = 2 $\quad$ $\ell = 2$ degen = 2 $\quad$ $\ell = 3$ degen = 2

energy eigenvalue

Problem

7.1 Normalization condition

What is the normalization condition for $u_{n,\ell}(r)$? Be sure to distinguish the cases $\ell = 0$ and $\ell \neq 0$.

7.2 Central potentials in three dimensions

The method used for central potentials in two dimensions works in three dimensions as well. The details are (as expected) messier: you have to use three spherical coordinates (r, θ, ϕ) rather than two polar coordinates (r, θ), so you have to use separation of variables with a product of three one-variable functions rather than a product of two one-variable functions. Thus there are two separation constants rather than one. Instead of presenting these messy details, I'll just quote the result:

To solve the three-dimensional energy eigenproblem for a spherically symmetric potential energy function $V(r)$, namely

$$-\frac{\hbar^2}{2M}\nabla^2\eta(\vec{r}) + V(r)\eta(\vec{r}) = E\eta(\vec{r}), \tag{7.28}$$

first solve the one-dimensional radial energy eigenproblem

$$-\frac{\hbar^2}{2M}\frac{d^2u(r)}{dr^2} + \left[V(r) + \frac{\hbar^2\ell(\ell+1)}{2M}\frac{1}{r^2}\right]u(r) = Eu(r) \tag{7.29}$$

for $\ell = 0, 1, 2, \ldots$. For a given ℓ, call the resulting energy eigenfunctions and eigenvalues $u_{n,\ell}(r)$ and $E_{n,\ell}$ for $n = 1, 2, 3, \ldots$. Then the three-dimensional solutions are

$$\eta_{n,\ell,m}(r, \theta, \phi) = \frac{u_{n,\ell}(r)}{r}Y_\ell^m(\theta, \phi) \quad \text{with energy } E_{n,\ell}, \tag{7.30}$$

where the "spherical harmonics" $Y_\ell^m(\theta, \phi)$ are particular special functions of the angular variables that you could look up if you needed to. The integer separation constant m takes on the $2\ell + 1$ values

$$-\ell,\ -\ell+1,\ \ldots,\ 0,\ \ldots,\ \ell-1,\ \ell.$$

Notice that the $2\ell + 1$ different solutions for a given n and ℓ, but with different m, are degenerate.

In addition, there's a strange terminology that you need to know. You'd think that the states with $\ell = 0$ would be called "$\ell = 0$ states", but in fact they're called "s states". You'd think that the states with $\ell = 1$ would be called "$\ell = 1$ states", but in fact they're called "p states". States with $\ell = 2$ are called "d states" and states with $\ell = 3$ are called "f states". (I am told that these names come from a now-obsolete system for categorizing atomic spectral lines as "sharp", "principal", "diffuse", and "fundamental". States with $\ell \geq 4$ are not frequently encountered, but they are called g, h, i, k, l, m, ... states. For some reason j is omitted. "Sober physicists don't find giraffes hiding in kitchens.")

In summary, the energy eigenvalues for some generic three-dimensional radially symmetric potential will look sort of like this:

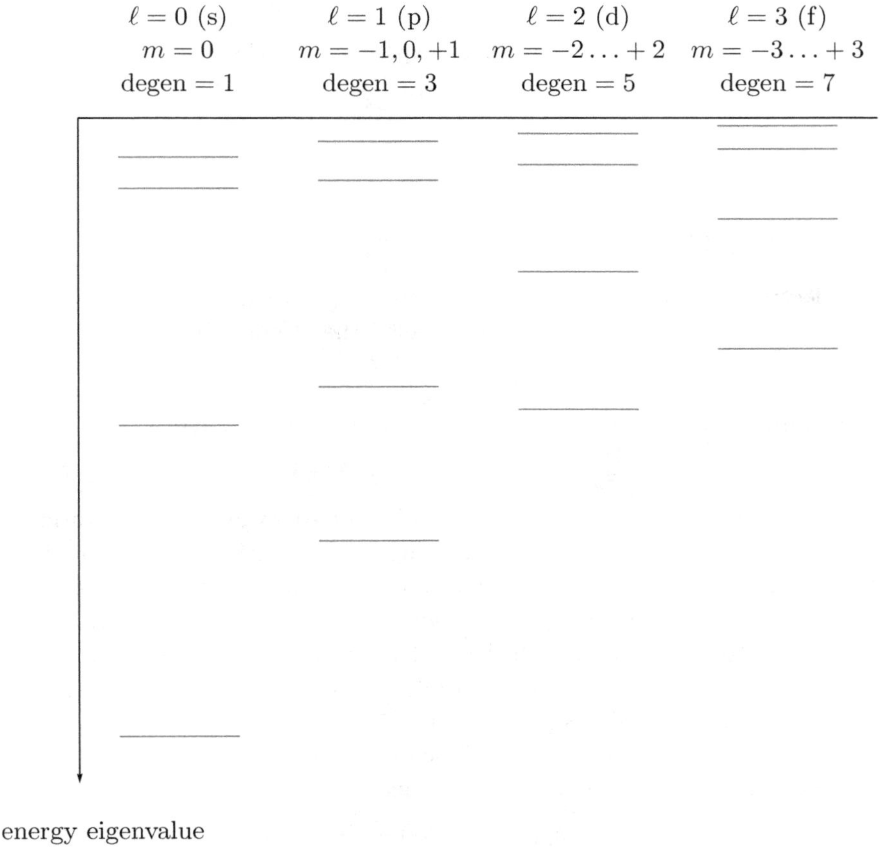

$\ell = 0$ (s)	$\ell = 1$ (p)	$\ell = 2$ (d)	$\ell = 3$ (f)
$m = 0$	$m = -1, 0, +1$	$m = -2 \ldots +2$	$m = -3 \ldots +3$
degen = 1	degen = 3	degen = 5	degen = 7

energy eigenvalue

This graph shows only the four lowest energy eigenvalues for each value of ℓ. A single horizontal line in the "$\ell = 0$ (s)" column represents a single energy eigenfunction, whereas a single horizontal line in the "$\ell = 2$ (d)" column represents five linearly independent energy eigenfunctions, each with the same energy ("degenerate states").

Problem

7.2 **Dimensions of $\eta(\vec{r})$ and of $u(r)$**
In equation (7.27) for the two-dimensional central potential problem, what are the dimensions of $\eta(\vec{r})$ and of $u(r)$? In equation (7.30) for the three-dimensional central potential problem, what are the dimensions of $\eta(\vec{r})$ and of $u(r)$? [This result helps motivate the definitions $u(r) = \sqrt{r}R(r)$ in two dimensions and $u(r) = rR(r)$ in three dimensions.]

7.3 The hydrogen atom

7.3.1 *The model*

An electron (of mass M) and a proton interact through the classical electrostatic potential energy function — called the "Coulomb potential" —

$$V(r) = -\frac{1}{4\pi\epsilon_0}\frac{e^2}{r}, \tag{7.31}$$

so you might think that the energy eigenproblem for the hydrogen atom is

$$-\frac{\hbar^2}{2M}\nabla^2\eta(\vec{r}) - \frac{1}{4\pi\epsilon_0}\frac{e^2}{r}\eta(\vec{r}) = E\eta(\vec{r}). \tag{7.32}$$

That's not exactly correct. This eigenproblem treats the proton as stationary while the electron does all the moving: in fact, although the proton is almost 2000 times more massive than the electron, it's not infinitely massive and it does do some moving. This eigenproblem assumes the proton is a point particle: in fact, although the nucleus is small compared to an atom, it does have some size. This eigenproblem is non-relativistic and it treats the electromagnetic field as purely classical: both false. This eigenproblem ignores the electron's spin. All of these are good approximations, but this is a *model* for a hydrogen atom, not the exact thing.[2]

But let's work with the approximation we have, rather than holding out for an exact solution of an exact eigenproblem that will never come.[3] What happens if we solve the three-dimensional central potential problem with the model potential energy function (7.31)? We don't yet have the mathematical tools to actually perform this solution, but we *are* in a position to appreciate the character of the solution.

[2]The corrections to the energy eigenvalues produced by equation (7.32) due to these effects are called "fine structure" and "hyperfine structure".

[3]Everyone knows that weather prediction is inexact. But you'd still rather know a prediction that a hurricane has an 80% chance of arriving at about 7:00 PM than be totally clueless about a hurricane bearing down on your home.

7.3.2 *The energy eigenvalues*

First of all, because the Coulomb potential *is* a particular kind of central potential, it will have all the properties listed in the last section for three-dimensional central potentials: Each energy eigenstate will be characterized by an ℓ and an m, where $\ell = 0, 1, 2, \ldots$ and where $m = -\ell, \ldots, +\ell$. The energy eigenvalues will be independent of m, resulting in degeneracy. And for a given n, the energy eigenvalue will increase with increasing ℓ.

The energy eigenvalues for the Coulomb potential turn out to be:

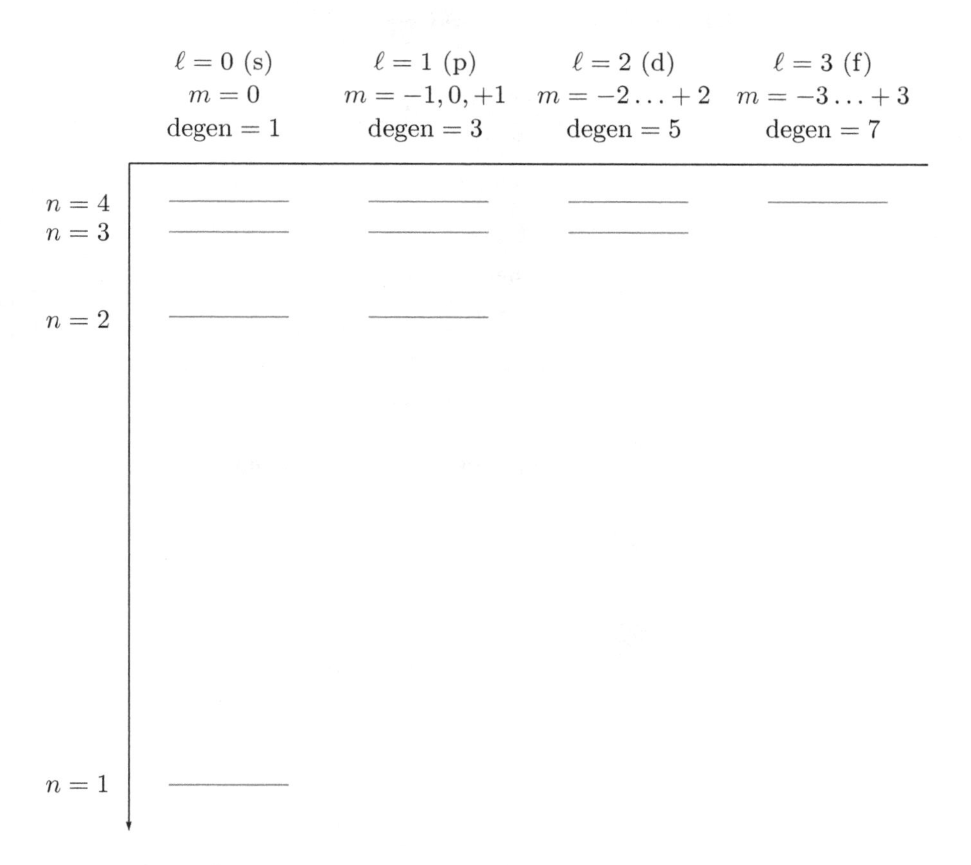

What a surprise! The energy eigenvalues for $\ell = 1$ are pushed up so much that they exactly line up with all but the lowest energy eigenvalues for $\ell = 0$. The energy eigenvalues for $\ell = 2$ are pushed up so much that they exactly line up with all but the lowest energy eigenvalues for $\ell = 1$. And so forth. This surprising line-up is called "accidental degeneracy".

Normally eigenfunctions are labeled by $n = 1, 2, 3, \ldots$. But this surprising line-up of energies suggests a different notation for the Coulomb problem. For "$\ell = 0$ (s)" the eigenfunctions are labeled as usual by $n = 1, 2, 3, 4, \ldots$. But for "$\ell = 1$ (p)" the eigenfunctions are labeled by $n = 2, 3, 4, \ldots$. For "$\ell = 2$ (d)" they are labeled by $n = 3, 4, \ldots$. And so forth. With this labeling scheme the energy eigenvalue turn out to be given by

$$E_n = -\frac{M}{2\hbar^2}\left(\frac{e^2}{4\pi\epsilon_0}\right)^2\frac{1}{n^2}. \tag{7.33}$$

The coefficient in this equation is called the "Rydberg[4] energy", and the equation is usually written

$$E_n = -\frac{\text{Ry}}{n^2}, \qquad \text{where} \qquad \text{Ry} = 13.6 \text{ eV}. \tag{7.34}$$

(I recommend that you memorize this energy 13.6 eV, the ionization energy for hydrogen, which sets the scale for typical energies in atomic physics. Much to my embarrassment, I forgot it during my graduate qualifying oral exam. Problem 7.4, "Dimensional analysis for energy eigenvalues", on page 245 presents a way to help understand and remember this result.)

[4]Johannes Rydberg (1854–1919), Swedish spectroscopist, discovered a closely related formula empirically in 1888. Do not confuse the Rydberg *energy* Ry = 13.6 eV with the Rydberg *constant* $R_\infty = 1.097 \times 10^7\ \text{m}^{-1}$.

7.3.3 *The energy eigenfunctions*

It's a triumph to know the energy eigenvalues, but we should know also something about the energy eigenfunctions, which are labeled $\eta_{n,\ell,m}(\vec{r})$. A terminology note is that an energy eigenfunction with $n = 3$, $\ell = 2$, and any value of m — that is $\eta_{3,2,m}(x)$ — is called a "3d state".

To gain this knowledge we need to first understand the effective potential energy function falling within square brackets in equation (7.29):

$$V_{\text{eff}}(r) = -\frac{1}{4\pi\epsilon_0}\frac{e^2}{r} + \frac{\hbar^2\ell(\ell+1)}{2M}\frac{1}{r^2}. \tag{7.35}$$

This function is sketched schematically on the next page. For large values of r, to the right in the sketch, $1/r$ is bigger than $1/r^2$, so $V_{\text{eff}}(r)$ is almost the same as the $1/r$ Coulomb potential energy alone. For small values of r, to the left in the sketch, $1/r$ is smaller than $1/r^2$, so $V_{\text{eff}}(r)$ is almost the same as the $1/r^2$ part alone. For intermediate values of r, the function $V_{\text{eff}}(r)$ has to swing between these two limits, as sketched.

The result of this swinging will of course depend upon the value of ℓ, and the results for four values of ℓ are sketched schematically on page 243.

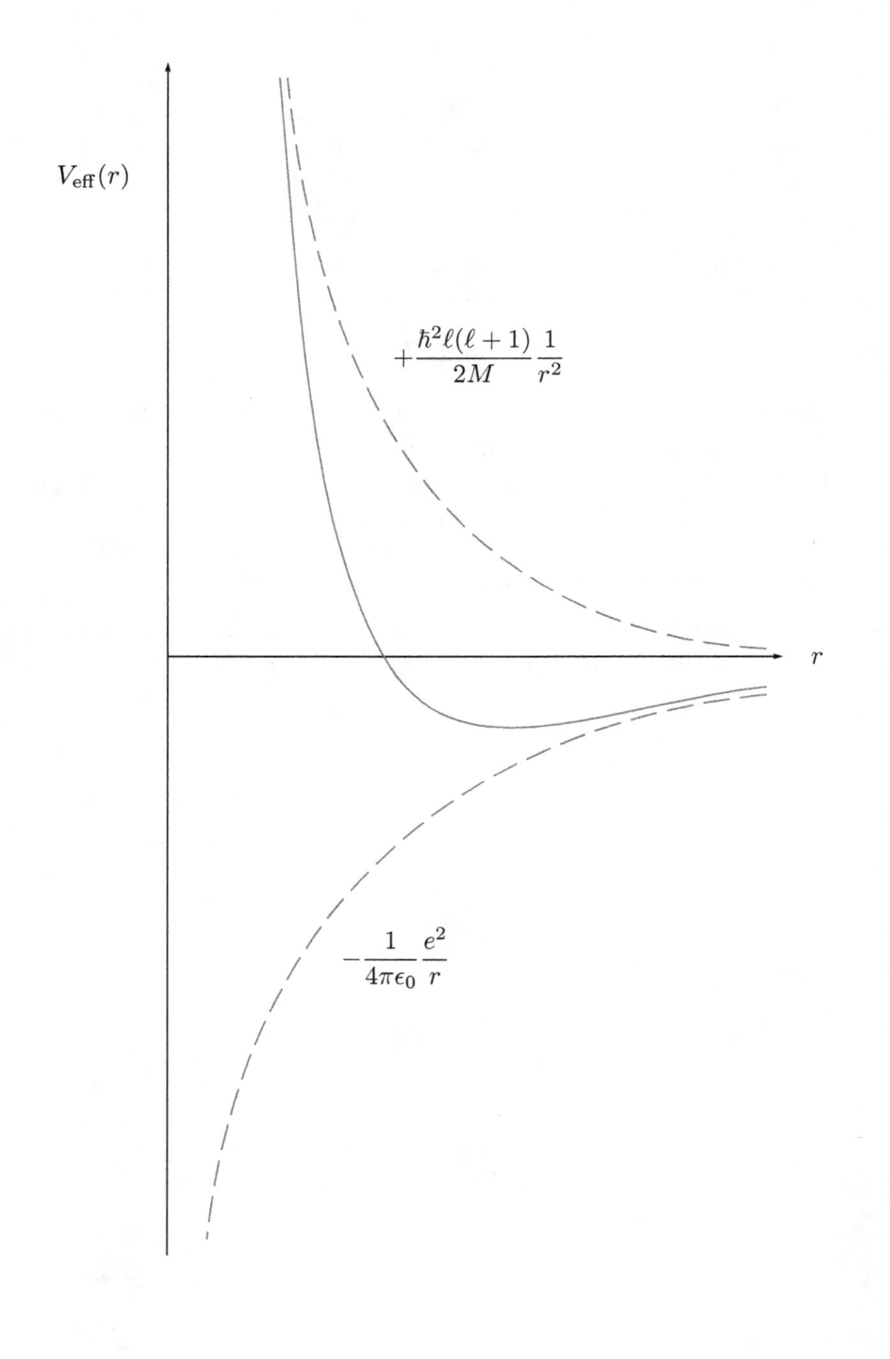
$V_{\text{eff}}(r)$
$+\frac{\hbar^2\ell(\ell+1)}{2M}\frac{1}{r^2}$
r
$-\frac{1}{4\pi\epsilon_0}\frac{e^2}{r}$

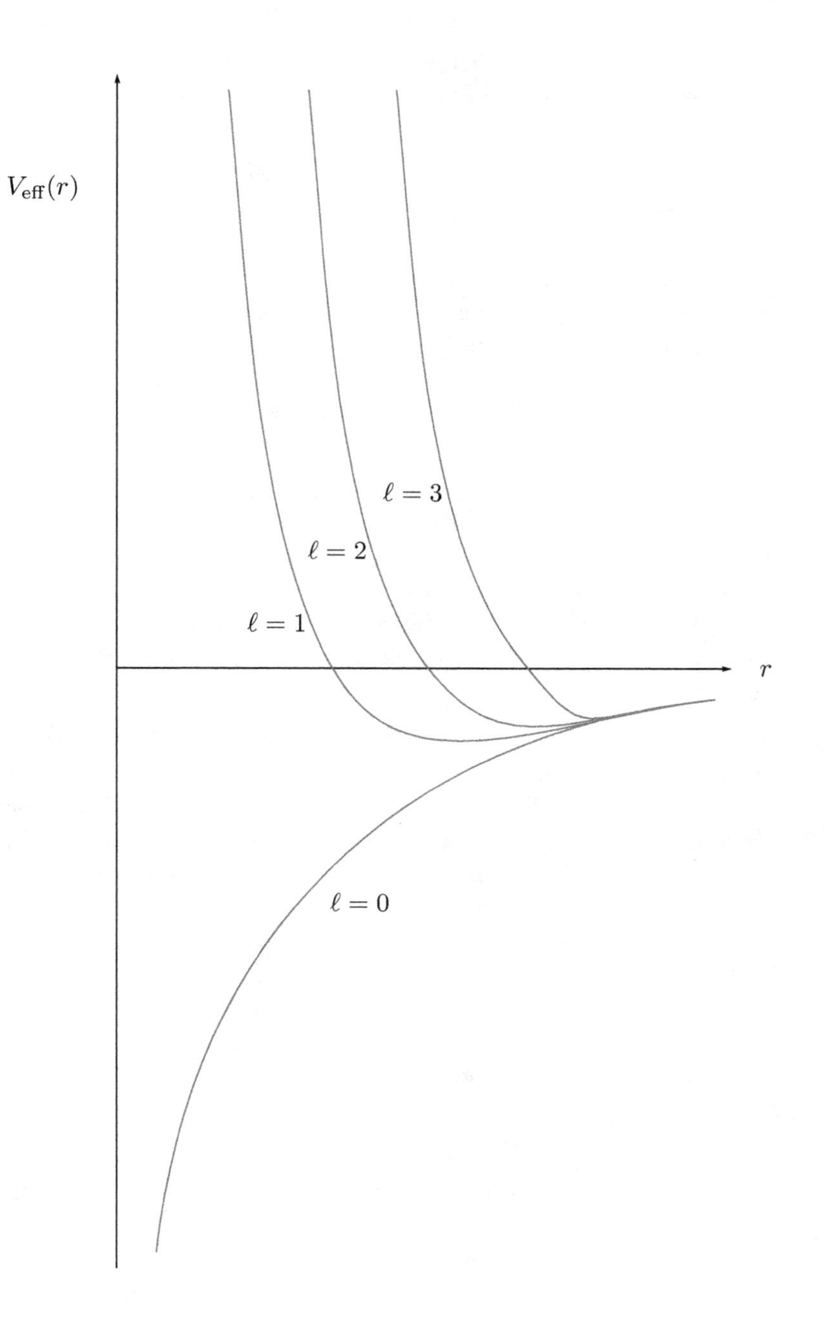
$V_{\text{eff}}(r)$
r
$\ell = 3$
$\ell = 2$
$\ell = 1$
$\ell = 0$

This graph suggests that for a given value of n, the states with larger ℓ will have larger mean values for r, the distance from the proton to the electron.

7.3.4 *Transitions*

If the energy eigenequation (7.32) for the hydrogen atom were exactly correct, then a hydrogen atom starting in the excited energy state $\eta_{3,2,-1}(\vec{r})$ would remain in that state forever. Furthermore, a hydrogen atom starting in a linear combination with probability 0.6 of being in energy state $\eta_{3,2,-1}(\vec{r})$ and probability 0.4 of being in energy state $\eta_{2,1,0}(\vec{r})$ would maintain those probabilities forever.

But the energy eigenequation (7.32) is *not* exactly correct. It ignores collisions, it ignores external electromagnetic field (e.g., incident light), and it ignores coupling to the electromagnetic field (e.g., radiated light). These effects mean that the state $\eta_{3,2,-1}(\vec{r})$ is a stationary state of the model eigenproblem, but it is not a stationary state of the exact eigenproblem. In other words, these effects result in transitions between stationary states of the model eigenproblem.

To understand these transitions you need to understand the transition-causing effects, and at this point in your education you're not ready to do that. But I'll tell you one thing right now: a transition involving a single photon (either absorbing or emitting a single photon) will result in a transition with $\Delta\ell = \pm 1$. So, for example, a hydrogen atom in a 2p state (that is, one with $n = 2$, $\ell = 1$, and any legal value of m) could transition to the 1s ground state by emitting a single photon. A hydrogen atom in a 2s state (that is, one with $n = 2$, $\ell = 0$, and $m = 0$) cannot transition to the ground state by emitting a single photon. (It could do so by emitting two photons, or through a collision.) An atom in the 1s ground state, exposed to a light source with photons of energy $\frac{3}{4}$Ry, can be excited to 2p state by absorbing a single photon, but it cannot be excited to the 2s state by absorbing a single photon.

I regard this fact (which, by the way, holds not only for the hydrogen atom but for any central potential) as a picky detail appropriate for an advanced course, but the people who write the Graduate Record Exam in physics seem to think it's important so you should probably remember it. (Or at least review this page the evening before you take the physics GRE.)

Problems

7.3 Counting hydrogen states

Show that the degeneracy for states characterized by n is n^2.

7.4 Dimensional analysis for energy eigenvalues

The eigenproblem (7.32) contains only two parameters:

$$\frac{\hbar^2}{M} \quad \text{and} \quad \frac{e^2}{4\pi\epsilon_0}.$$

Use dimensional analysis to show that these two parameters can come together to form an energy in only one way. ⟦I remember the Rydberg energy as

$$\text{Ry} = \frac{1}{2}\frac{(e^2/4\pi\epsilon_0)^2}{\hbar^2/M} \tag{7.36}$$

using this dimensional analysis trick.⟧

7.5 Which states are distant, which are close? (essential problem)

Argue, on the basis of the graph on page 242, that for a given value of ℓ, states with larger n will have larger mean values for r.

7.6 Energy eigenvalues for the He$^+$ ion (essential problem)

A helium atom with one electron stripped away is called a He$^+$ ion. This situation is simply one electron ambivating in the potential energy established by a highly-massive nucleus: it is just like the hydrogen atom, except that the nuclear change is $+2e$ rather than $+e$. At the level of approximation used in equation (7.32), the energy eigenproblem for the He$^+$ ion is

$$-\frac{\hbar^2}{2M}\nabla^2\eta(\vec{r}) - \frac{1}{4\pi\epsilon_0}\frac{2e^2}{r}\eta(\vec{r}) = E\eta(\vec{r}). \tag{7.37}$$

Show that (at this level of approximation) the energy eigenvalues for the He$^+$ ion are

$$E_n = -4\frac{\text{Ry}}{n^2}. \tag{7.38}$$

The situation of a single electron ambivating in the potential energy established by a highly-massive carbon nucleus of charge $+6e$ is called the C^{5+} ion. Show that (at this level of approximation) the energy eigenvalues are

$$E_n = -36\frac{\text{Ry}}{n^2}. \tag{7.39}$$

7.7 **Characteristic quantities for the Coulomb problem**

The time evolution Schrödinger equation for the Coulomb problem is

$$\frac{\partial\psi(\vec{r},t)}{\partial t} = -\frac{i}{\hbar}\left[-\frac{\hbar^2}{2M}\nabla^2\psi(\vec{r},t) - \frac{e^2}{4\pi\epsilon_0}\frac{1}{r}\psi(\vec{r},t)\right]. \qquad (7.40)$$

There are only three parameters in this equation: $\hbar$, M, and $e^2/4\pi\epsilon_0$. Using the techniques of sample problem 5.2.1 on page 200, find the characteristic time and length for the Coulomb problem. Define the scaled quantities

$$\tilde{t} = \frac{t}{\text{characteristic time}} \quad \text{and} \quad \tilde{\vec{r}} = \frac{\vec{r}}{\text{characteristic length}},$$

and write the time evolution equation (7.40) in terms of these variables. If you didn't like to explain what you were doing (or if you wanted to sound cryptic to impress the uninitiated) how would you use shorthand to describe the result of this scaling strategy?

7.8 **Hybridization**

For some chemical applications, it is useful to define the four "sp^3 hybrid states"

$$\begin{aligned}
\phi_1(\vec{r}) &= \tfrac{1}{2}\left[\eta_{2,0,0}(\vec{r}) + \eta_{2,1,+1}(\vec{r}) + \eta_{2,1,0}(\vec{r}) + \eta_{2,1,-1}(\vec{r})\right] \\
\phi_2(\vec{r}) &= \tfrac{1}{2}\left[\eta_{2,0,0}(\vec{r}) + \eta_{2,1,+1}(\vec{r}) - \eta_{2,1,0}(\vec{r}) - \eta_{2,1,-1}(\vec{r})\right] \\
\phi_3(\vec{r}) &= \tfrac{1}{2}\left[\eta_{2,0,0}(\vec{r}) - \eta_{2,1,+1}(\vec{r}) - \eta_{2,1,0}(\vec{r}) + \eta_{2,1,-1}(\vec{r})\right] \\
\phi_4(\vec{r}) &= \tfrac{1}{2}\left[\eta_{2,0,0}(\vec{r}) - \eta_{2,1,+1}(\vec{r}) + \eta_{2,1,0}(\vec{r}) - \eta_{2,1,-1}(\vec{r})\right].
\end{aligned}$$

a. Which, if any, of these are energy eigenstates? What is the energy eigenvalue associated with each such eigenstate?

b. The eigenstates $\eta_{n,\ell,m}(\vec{r})$ are "orthonormal" in the sense that the integral over all space satisfies

$$\int \eta^*_{n',\ell',m'}(\vec{r})\eta_{n,\ell,m}(\vec{r})\, d^3r = \begin{cases} 1 \text{ if } n' = n,\ \ell' = \ell, \text{ and } m' = m \\ 0 \text{ otherwise} \end{cases}. \qquad (7.41)$$

Evaluate the sixteen integrals

$$\int \phi_i^*(\vec{r})\phi_j(\vec{r})\, d^3r. \qquad (7.42)$$

Clue: This is not a difficult problem. If you're working hard, then you're working too hard.

7.4 The helium atom

Here's the energy eigenproblem for the helium atom, at the same level of approximation as the eigenproblem (7.32) for the hydrogen problem:

$$\begin{aligned}
&-\frac{\hbar^2}{2M}\nabla_A^2\eta(\vec{r}_A,\vec{r}_B)-\frac{\hbar^2}{2M}\nabla_B^2\eta(\vec{r}_A,\vec{r}_B)\\
&-\frac{1}{4\pi\epsilon_0}\frac{2e^2}{r_A}\eta(\vec{r}_A,\vec{r}_B)-\frac{1}{4\pi\epsilon_0}\frac{2e^2}{r_B}\eta(\vec{r}_A,\vec{r}_B)+\frac{1}{4\pi\epsilon_0}\frac{e^2}{|\vec{r}_A-\vec{r}_B|}\eta(\vec{r}_A,\vec{r}_B)\\
&\quad=E\eta(\vec{r}_A,\vec{r}_B). \qquad (7.43)
\end{aligned}$$

We have no chance whatsoever of solving this "two electron plus one nucleus" problem exactly. Even the classical problem of three particles interacting through $1/r$ potentials, first posed by Isaac Newton in 1687, has not yet been solved exactly. (And probably never will be, because the resulting behavior is known to be chaotic.) Since classical mechanics is a subset of quantum mechanics, an exact solution to this helium problem would contain within it an exact solution to the unsolved classical "three-body problem".

Does this mean we should give up? Not at all. We should instead look for approximate solutions that are not exact, but highly accurate for the bound-state regime of interest.

Our approach will involve solving the one-electron problem for a different potential, and then using those one-electron levels as building blocks for the two-electron problem through the antisymmetrization machinery of equation (6.11). The strategy may seem crude, but in practice it can produce highly accurate results.

Instead of focusing on two electrons, interacting with the nucleus and with each other, focus on one electron interacting with the nucleus and with the average of the other electron. I don't yet know exactly how the "other" electron is averaged, but I assume it spreads out in a spherically symmetric cloud-like fashion.

Finding the potential. Remember, from your electrostatics course, the shell theorem for spherically symmetric charge distributions: When the electron under focus is close to the nucleus, it feels only the electric field due to the nucleus, so the potential energy is

$$\text{for small } r, \quad V(r)\approx-\frac{1}{4\pi\epsilon_0}\frac{2e^2}{r}. \qquad (7.44)$$

Whereas when the electron under focus is far from the nucleus, it feels the electric field due to the nucleus, plus the electric field due to the cloud collapsed into the nucleus, so the potential energy is

$$\text{for large } r, \qquad V(r) \approx -\frac{1}{4\pi\epsilon_0}\frac{e^2}{r}. \tag{7.45}$$

The potential energy felt by the electron under focus will interpolate between these two limits, something like the solid line graphed below.

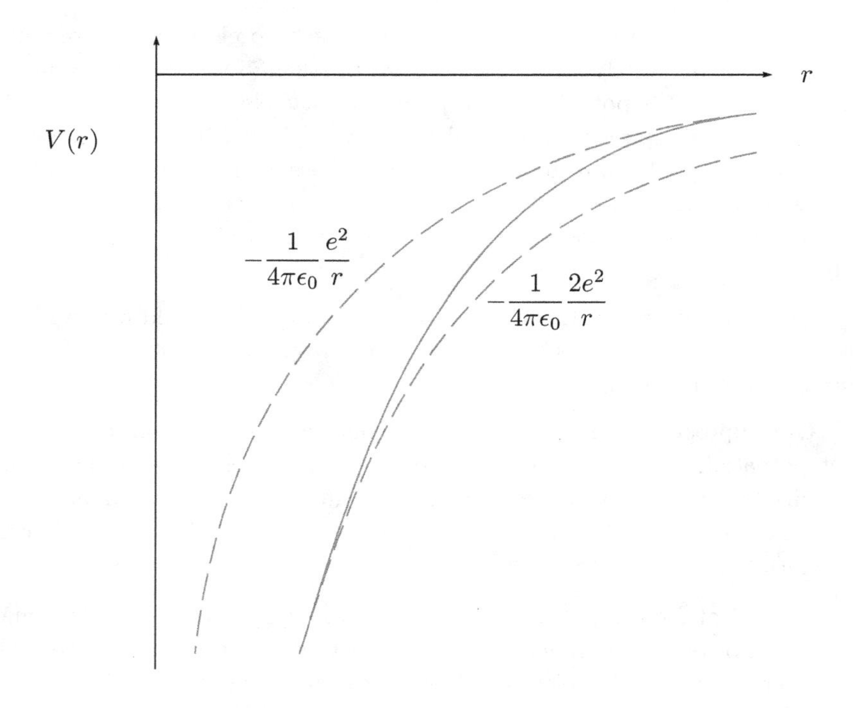

This phenomenon is called "shielding". The shielded potential interpolates between the known limits of small r and large r behavior. The exact character of that interpolation is unclear: if you were doing high-accuracy calculations, you would need to find it.[5] For our purposes it will be enough just to know the two limits.

[5]Using a technique called the Hartree-Fock approximation.

This approach is called a "mean field approximation" and is said to "ignore correlations" between the two electrons. This approach is not exact and cannot be made exact, but it enables progress to be made.

Finding the one-electron eigenvalues. What will the one-electron energy eigenvalues be for a shielded potential energy? If the potential energy were

$$V(r) = -\frac{1}{4\pi\epsilon_0}\frac{e^2}{r}, \tag{7.46}$$

then the system would be a hydrogen atom (H atom) and the energy eigenvalues would be

$$E_n = -\frac{\text{Ry}}{n^2}. \tag{7.47}$$

If the potential energy were

$$V(r) = -\frac{1}{4\pi\epsilon_0}\frac{2e^2}{r}, \tag{7.48}$$

then the system would be a positively charged helium ion (He^+ ion; a helium atom with one electron stripped away); equation (7.38) shows that the energy eigenvalues would be

$$E_n = -4\frac{\text{Ry}}{n^2}. \tag{7.49}$$

But in fact, the potential energy interpolates between these two forms, so the energy eigenvalues interpolate between these two possibilities. Let's examine this interpolation, first for the s levels:

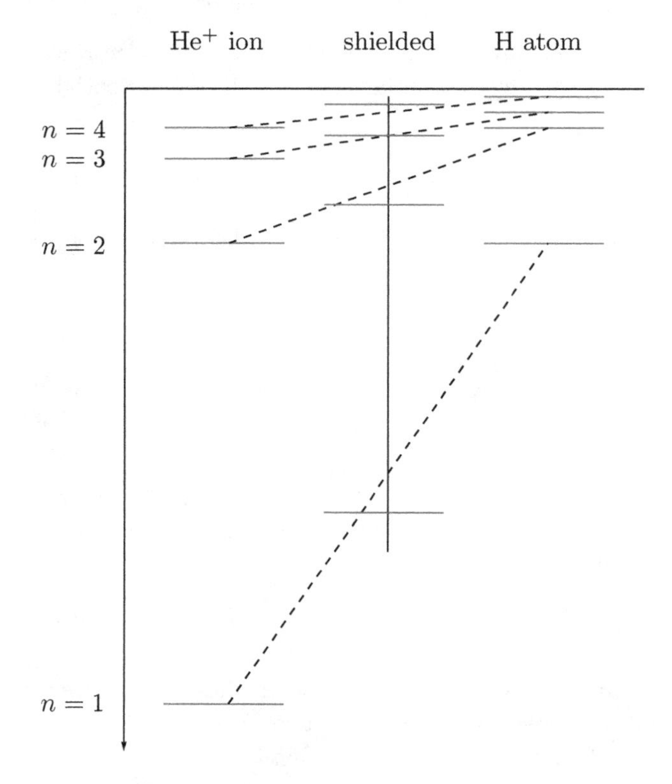

To the left, under the heading "He^+ ion", are the s state eigenvalues (7.49), which are four times deeper than those under the heading "H atom", the s state eigenvalues (7.47). I've drawn a vertical line midway between them and dashed lines connecting the the two sets of eigenvalues. If the eigenvalues for the shielded potential were exactly halfway between the eigenvalues for the two limits, then they would fall where the dashed lines cross the vertical line. But they don't fall exactly there. For states that are mostly near the nucleus, the energies are closer to (7.49). For states that are mostly far from the nucleus, the energies are closer to (7.47). We have already seen (problem 7.5 on page 245) that, for a given ℓ, the eigenfunctions with larger

n are farther from the nucleus. Chemists like to say that the eigenfunctions that are mostly close to the nucleus — those with smaller n — have more "penetration".

This process can be repeated for p states, d states, and f states. Because, for a given n, the eigenfunction with larger ℓ is farther from the nucleus (less "penetration", see page 244), the eigenfunction with larger ℓ will have higher energy. Thus a shielded potential energy function will give rise to a set of energy eigenvalues like this:

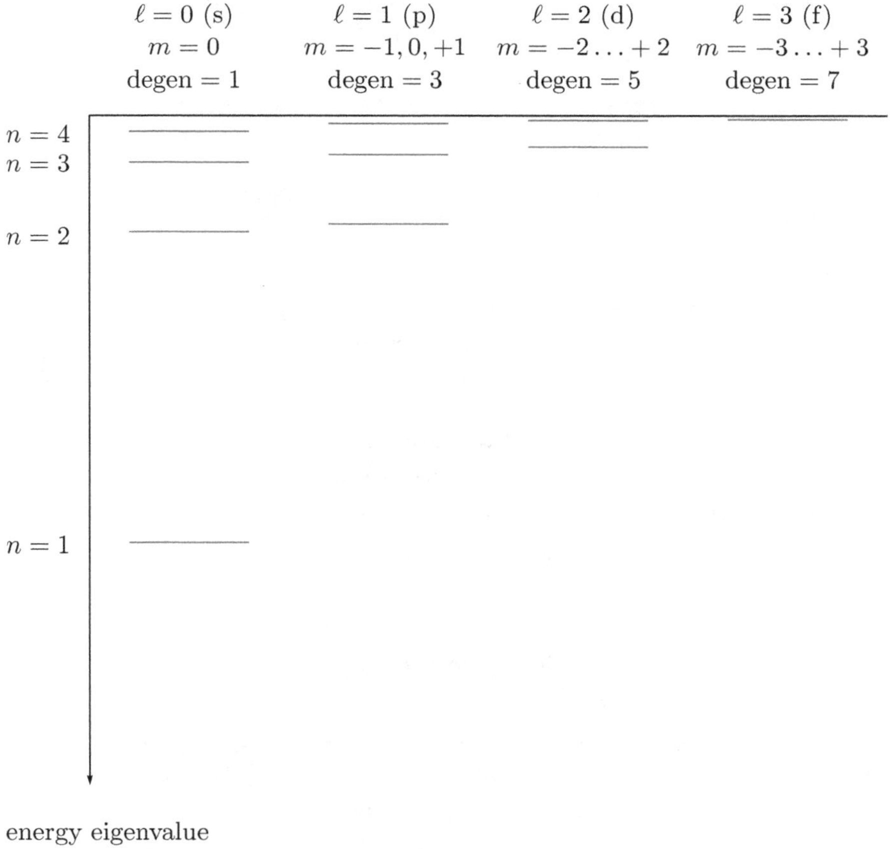

$\ell = 0$ (s)	$\ell = 1$ (p)	$\ell = 2$ (d)	$\ell = 3$ (f)
$m = 0$	$m = -1, 0, +1$	$m = -2 \ldots + 2$	$m = -3 \ldots + 3$
degen = 1	degen = 3	degen = 5	degen = 7

$n = 4$

$n = 3$

$n = 2$

$n = 1$

energy eigenvalue

Building two-electron states from one-electron levels. Now that we have one-electron eigenfunctions (the "levels"), we can combine them through the antisymmetrization machinery to produce two-electron eigenfunctions (the "states"). This process was described in section 6.6 on page 219.

Misconception. Perhaps you learned in high-school chemistry that the ground state of the helium atom has "one electron in the '1s, spin up' level, and one electron in the '1s, spin down' level". That's just plain wrong — if it were right, then you'd be able to distinguish between the two electrons (the one with spin up, the one with spin down) and then the two electrons wouldn't be identical. What's correct is that the *individual* electrons don't have states: instead the *pair* of electrons is in one state, namely the antisymmetric non-product state

$$\eta_{1,0,0}(\vec{x}_A)\eta_{1,0,0}(\vec{x}_B)\frac{1}{\sqrt{2}}[\chi_+(A)\chi_-(B) - \chi_-(A)\chi_+(B)].$$

(Compare problem 6.10, "Intersystem crossing", on page 225.)

Problem

7.9 **The hydrogen molecule ion**

If a hydrogen molecule H_2 is stripped of one electron, the result is the "hydrogen molecule ion" consisting of two nuclei and one electron. Show that if we had solved the helium problem exactly we would have also solved the hydrogen molecule ion problem. (But we have not solved the problem exactly: instead we found approximations appropriate for the case of atomic helium. A completely different set of approximations are appropriate for the hydrogen molecule ion.)

7.5 The lithium atom

This situation (within the shielded potential approximation) was discussed in section 6.7 on page 225. In summary:

The ground state of hydrogen is two-fold degenerate. The ground state of helium is non-degenerate. The ground state of lithium is (within the shielded potential approximation) two-fold degenerate.

The ground states of hydrogen and helium can (within the shielded potential approximation) be written as a spatial part times a spin part. The ground state of lithium cannot be so written.

7.6 All other atoms

Atoms larger than lithium are difficult. Specialists have examined them in exquisite detail, but in this book we're not going to try to find the energy spectrum, we're not going to try to find the ground state degeneracy, we're not even going to try to write down a ground state. Instead, we're only going to list the one-electron levels that are thrown together through the antisymmetrization machinery (6.11) to make the many-electron ground state.

Try this buliding-up machinery for carbon, with a nucleus and six electrons.[6] In the figure on page 250, pertaining to helium, the left-hand energies are four times deeper than the right-hand energies. If I were to draw a parallel figure for carbon (see equation 7.39), the left-hand energies would be 36 times deeper than the right-hand energies! The net result is that, while the figure on page 251 shows a modest increase in energy $E_{n,\ell}$ for a given n as you move right to higher values of ℓ, for carbon the energy increase will be dramatic: something like the figure on the next page. For atoms bigger than carbon, the increase will be still more dramatic.

[6]Carbon is a seven-body problem, not a three-body problem like helium, so of course it is correspondingly less tractable.

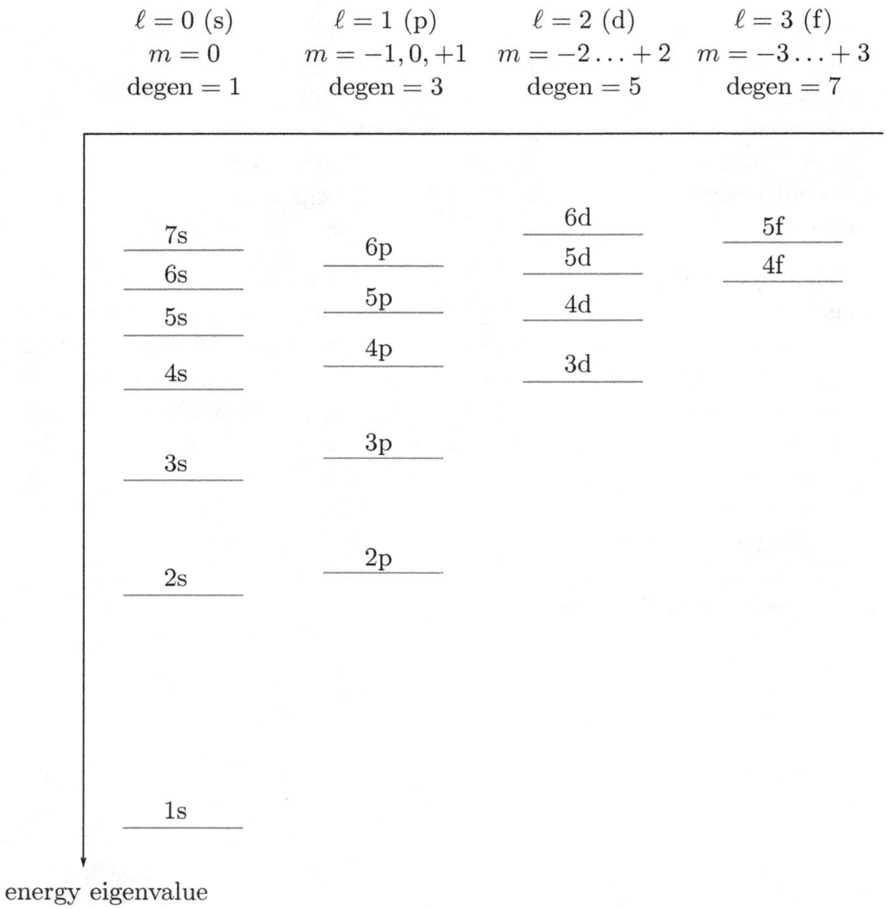

$\ell = 0$ (s)	$\ell = 1$ (p)	$\ell = 2$ (d)	$\ell = 3$ (f)
$m = 0$	$m = -1, 0, +1$	$m = -2 \ldots +2$	$m = -3 \ldots +3$
degen = 1	degen = 3	degen = 5	degen = 7
7s	6p	6d	5f
6s	5p	5d	4f
5s	4p	4d	
4s	3p	3d	
3s	2p		
2s			
1s			

energy eigenvalue

Figure 7.1: *Schematic energy levels for a generic atom, about the size of carbon.*

I underscore once more that these levels come from an approximation that ignores relativity, electron spin, nuclear size, nuclear motion, and the quantal character of the electromagnetic field. Most damning of all, it replaces electron-electron repulsion with a shielded potential. Since this shielded potential is spherically symmetric, the language of ℓ and m and 2p levels and so forth can be used. But this is only an approximation.

With this understanding, we ask which one-electron levels will go into the antisymmetrization machinery to make the carbon six-electron ground state. The ground state will be constructed from the six lowest possible energy levels. There will be a "1s, spin up" level, plus a "1s, spin down" level, plus a "2s, spin up" level, plus a "2s, spin down" level. In addition there will be two 2p levels, but it's not clear which ones they will be: Will they be the 2p level with $m = 0$ and spin up, plus the 2p level with $m = +1$ and spin up? Or will they be the 2p level with $m = 0$ and spin up, plus the 2p level with $m = 0$ and spin down? Or will they be some other superposition?

At the level of approximation used here, all such combinations have exactly the same energy: they are degenerate. If you study more atomic physics you'll learn Hund's[7] rules for figuring out how the degeneracy is broken at a more accurate level of approximation. But for the purposes of this book, it's only necessary to list the ns and the ℓs of the one-electron levels that go into making the six-electron ground state. This list is called the "electronic configuration" and it's represented through a special notation: the configuration of carbon is written $1s^2 2s^2 2p^2$.

For still larger atoms, the shielding effect is more dramatic and the energy levels shift still further, but still, usually, the one-electron levels fall within the energy sequence shown on page 254. This is the so-called "Madelung[8] sequence" of level energies:

$$
\begin{aligned}
&1s < 2s < 2p < 3s < 3p < 4s < 3d < 4p \\
&\quad < 5s < 4d < 5p < 6s < 4f < 5d < 6p < 7s < 5f < 6d \ \ldots
\end{aligned}
$$

It would be a miracle indeed if the theoretical calculations for *all* the different atoms resulted in *exactly* the same qualitative sequence of energy levels. And it would be more miraculous still if the approximations used were close enough to reality that the prediction of the approximation was always accurate. And indeed neither of these miracles occur.[9] For example in chromium, atomic number 24, the configuration predicted by the

[7]Friedrich Hund (1896–1997), German physicist who applied quantum mechanics to atoms and molecules, and who discovered quantum tunneling.

[8]Erwin Madelung (1881–1972), German physicist with interests in crystal structure, atomic physics, and quantum mechanics. He produced a set of equations equivalent to the Schrödinger equation but which emphasized the flow of probability density rather than of amplitude density.

[9]See W.H. Eugen Schwarz and Ronald L. Rich, "Theoretical basis and correct explanation of the periodic system: Review and update" *Journal of Chemical Education* **87** (April 2020) 435–443; Gregory Anderson, Ravi Gomatam, and Laxmidhar Behera,

Madelung sequence is $1s^2 2s^2 2p^6 3s^2 3p^6 4s^2 3d^4$ whereas experiment shows the actual configuration ends instead with $4s^1 3d^5$.

Problem

7.10 **Ground state degeneracy**
At the level of approximation of the diagram on page 254, find the degeneracy of the ground state of boron, of carbon, of nitrogen, of oxygen, of fluorine, and of neon.

7.7 The periodic table

While the Madelung sequence is not perfect (few things are), it makes sense to see what it has to say ("better to light a single candle, no matter how faint, than to curse the darkness").

Compare the levels that go into building up carbon (atomic number 6) with those that go into building up silicon (atomic number 14). For carbon they are $1s^2 2s^2 2p^2$; for silicon they are $1s^2 2s^2 2p^6 3s^2 3p^2$. Note the similarities of those last, highest energy levels: carbon ends with $2s^2 2p^2$, silicon ends with $3s^2 3p^2$. It's possible that for carbon the nucleus (charge $+6$) and the 1s and 2s electron levels (charge -4) act together to form an atom core of net charge $+2$. Meanwhile it's just as possible that for silicon the nucleus (charge $+14$) and the 1s, 2s, 2p, and 3s electron levels (charge -12) similarly act together to form an atom core again with net charge $+2$. If this possibility is correct, then you would expect carbon and silicon to have similar chemical behavior. Sure enough each of them bonds with four other atoms: methane (CH_4) is chemically analogous to silane (SiH_4).

Through parallel reasoning you would expect neon, $1s^2 2s^2 2p^6$, to behave similarly to argon, $1s^2 2s^2 2p^6 3s^2 3p^6$. Again this expectation holds: both neon and argon are noble gases that react reluctantly with other atoms. Chemists have an ingenious system for showing these chemical similarities through a graphic called "the periodic table", shown on the next page.

"Contradictions in the quantum mechanical explanation of the periodic table" *Journal of Physics: Conference Series* **490** (2014) 012197.

1	H 1																		He 2
2	Li 3	Be 4												B 5	C 6	N 7	O 8	F 9	Ne 10
3	Na 11	Mg 12												Al 13	Si 14	P 15	S 16	Cl 17	Ar 18
4	K 19	Ca 20		Sc 21	Ti 22	V 23	Cr 24	Mn 25	Fe 26	Co 27	Ni 28	Cu 29	Zn 30	Ga 31	Ge 32	As 33	Se 34	Br 35	Kr 36
5	Rb 37	Sr 38		Y 39	Zr 40	Nb 41	Mo 42	Tc 43	Ru 44	Rh 45	Pd 46	Ag 47	Cd 48	In 49	Sn 50	Sb 51	Te 52	I 53	Xe 54
6	Cs 55	Ba 56	*	Lu 71	Hf 72	Ta 73	W 74	Re 75	Os 76	Ir 77	Pt 78	Au 79	Hg 80	Tl 81	Pb 82	Bi 83	Po 84	At 85	Rn 86
7	Fr 87	Ra 88	**	Lr 103	Rf 104	Db 105	Sg 106	Bh 107	Hs 108	Mt 109	Ds 110	Rg 111	Cn 112						

*	La 57	Ce 58	Pr 59	Nd 60	Pm 61	Sm 62	Eu 63	Gd 64	Tb 65	Dy 66	Ho 67	Er 68	Tm 69	Yb 70
**	Ac 89	Th 90	Pa 91	U 92	Np 93	Pu 94	Am 95	Cm 96	Bk 97	Cf 98	Es 99	Fm 100	Md 101	No 102

Let's carry out this building-up scheme systematically. Hydrogen and helium we have already discussed. Lithium adds one 2s level to the mix,[10] and beryllium one more. The next six elements each add one 2p level, ending at neon. Now pass on to row 3 of the periodic table. Sodium (atomic number 11, symbol Na from the Latin "natrium") has one 3s level just as lithium has one 2s level. Sure enough lithium and sodium are both highly reactive, chemically, and they react in similar ways. (They are called "alkali metals".) As we add electron levels, the march right on row 3 parallels the march right on row 2, ending in the noble gas argon, atomic number 18.

Row 4 starts out just like rows 2 and 3: potassium (atomic number 19, symbol K from the Latin "kalium") is chemically similar to lithium and sodium; calcium is chemically similar to beryllium and magnesium. But scandium, atomic number 21, is not at all like boron, atomic number 5. That's because the highest energy level in boron is a 2p level. The highest

[10]To write this out in detail, the three-electron wavefunction for the lithium ground state comes from feeding a 1s spin up level, plus a 1s spin down level, plus now a 2s level, into the "multiply and antisymmetrize" machinery at equation (6.11) in order to generate a three-electron state. This is a real mouthful, so we simply write "add one 2s level". In some books you will see this written as "add one 2s electron", but that language reinforces the misconception that the electron in the 2s level can be distinguished from the other two electrons, in which case the three electrons would not be identical. See the warnings on page 252 and in problem 6.10, "Intersystem crossing", on page 225.

energy level in scandium is not a p level at all, it's a 3d level. There are ten such levels, accounting for the ten elements from scandium through zinc. The element beyond zinc is gallium, which sure enough shares a lot of properties with its vertical neighbor aluminum. Row 4 continues by adding 4p levels until ending with the noble gas krypton.

Row 5 is very much like row 4.

Row 6 adds a new twist after barium, atomic number 56. The extra level in lanthanum, atomic number 57, is a 4f level. The fourteen 4f levels account for the fourteen elements from lanthanum through ytterbium. (The page in this book is not wide enough to hold row 6 of the periodic table, so I shoehorn in these fourteen elements using an asterisk.) Then the ten 5d levels account for the ten elements from lutetium through mercury (symbol Hg from the Greek "hydrargyrum", meaning "liquid silver"). Finally the six 6p levels account for the six elements from thallium through the noble gas radon.

Row 7 is very much like row 6. Many of the elements in row 7 have short-lived nuclei, but that's a different story.

Any chemist will tell you, correctly, that this lightning tour of the periodic table leaves out a lot of fascinating detail. But its very briefness means that you have not been distracted by detail and have kept sight of the central fact that the entire structure of the periodic table — and hence all of chemistry — follows from the Pauli requirement for antisymmetry under fermion coordinate swaps.

Problem

7.11 **Questions** (recommended problem)
Update your list of quantum mechanics questions that you started at problem 1.17 on page 46. Write down new questions and, if you have uncovered answers to any of your old questions, write them down briefly.

⟦For example, one of my questions would be: "The text claims on page 256 that the experimentally determined configuration for chromium, element 24, is $1s^2 2s^2 2p^6 3s^2 3p^6 4s^1 3d^5$. How can experiment determine such a thing?"⟧

Chapter 8

The Vistas Open to Us

I reckon I got to light out for the territory ahead...
— Mark Twain (last sentence of *Huckleberry Finn*)

This is the last chapter of the book, but this book itself is an invitation only, so it is not the last chapter of quantum mechanics. There are many fascinating topics that this book hasn't even touched on. Quantum mechanics will — if you allow it — surprise and delight and mystify you for the rest of your life.

This book devotes two chapters to **qubits**, also called spin-$\frac{1}{2}$ systems. Plenty remains to investigate: "which path" interference experiments, delayed-choice interference experiments, many different entanglement situations. For example, we developed entanglement through a situation where the quantal probability was $\frac{1}{2}$ while the local deterministic probability was $\frac{5}{9}$ or more (page 80). Different, to be sure, but not dramatically different. In the Greenberger–Horne–Zeilinger entanglement situation the quantal probability is 1 and the local deterministic probability is 0. You can't find probabilities more different than that! If you find these situations as fascinating as I do, then I recommend George Greenstein and Arthur G. Zajonc, *The Quantum Challenge: Modern Research on the Foundations of Quantum Mechanics.*

For many decades, research into qubits yielded insight and understanding, but no practical applications. All that changed with the advent of quantum computing. This is a rapidly changing field, but the essay "Quantum Entanglement: A Modern Perspective" by Barbara M. Terhal, Michael M. Wolf, and Andrew C. Doherty (*Physics Today*, April 2003) contains core insights that will outlive any transient. From the abstract: “It’s not your grandfather’s quantum mechanics. Today, researchers treat entanglement as a physical resource: Quantum information can now be measured, mixed, distilled, concentrated, and diluted.”

Because quantum mechanics is both intricate and unfamiliar, a formidable yet beautiful mathematical **formalism** has developed around it: position wavefunctions, momentum wavefunctions, Fourier transforms, operators, Wigner functions. These are powerful precision tools, so magnificent that some confuse the tools with nature itself. Every quantum mechanics textbook develops this formalism to a greater or lesser extent, but I also recommend the cute book by Leonard Susskind and Art Friedman, *Quantum Mechanics: The Theoretical Minimum.*

Here are two formal problems to whet your appetite: Back on page 140 I quoted O. Graham Sutton that “A technique succeeds in mathematical physics, not by a clever trick, or a happy accident, but because it expresses some aspect of a physical truth.” Thus inspired, we asked about the meaning of the separation constant (4.16), and that inquiry led us to the whole structure of stationary states and the energy eigenproblem. (And to a poem by T.S. Eliot, page 146.) But when we faced a similar separation constant at equation (7.10) we were too busy to follow up and ask what it was telling us. If you study more quantum mechanics, you will learn that this separation constant is related to angular momentum, that angular momentum is related to rotations, and that the conservation of angular momentum is related to rotational symmetry!

The second problem involves the simple harmonic oscillator, that is, the potential energy function $V(x) = \frac{1}{2}kx^2$. As with any one-dimensional potential well there are energy eigenstates. If any one of these states is shifted by a distance, it is of course no longer an energy eigenstate, so it *does* change with time. The remarkable thing is *how* it changes with time: the probability density does not spread, nor compact, nor change shape. Instead it rigidly slides back and forth with the same period that a classical particle would have in that same potential well.[1] When I first did the math to show that this is so, I was so astounded that I wrote a computer program to check out the math. This remarkable fact is true, and I have the feeling

[1]M.E. Marhic, “Oscillating Hermite-Gaussian wave functions of the harmonic oscillator” *Lettere al Nuovo Cimento* **22** (1978) 376–378, and C.C. Yan, “Soliton like solutions of the Schrödinger equation for simple harmonic oscillator” *American Journal of Physics* **62** (1994) 147–151.

that it "expresses some aspect of a physical truth", but I have no idea of what that physical truth might be.

We have applied quantum mechanics to cryptography, to model systems, and to atoms. **Applications** continue to molecules and to solids, to nuclei and to elementary particles, to superfluids, superconductors, and lasers, to liquid crystals, polymers, and membranes; the list is endless. Indeed, sunlight itself is generated through a quantal tunneling process! White dwarf stars work because of quantum mechanics, so do transistors and light-emitting diodes. In 1995 a new state of matter, the Bose-Einstein condensate, came into existence in a laboratory in Boulder, Colorado. In 2003 an even more delicate state, the fermionic condensate, was produced, again in Boulder. Both of these states of matter exist because of the Pauli principle, applied over and over again to millions of atoms.

Way back on page 5 we mentioned the need for a **relativistic quantum mechanics** and its associate, quantum field theory. The big surprise is that these theories don't just treat particles moving from place to place. They predict that particles can be created and destroyed, and sure enough that happens in nature under appropriate conditions.

There's **plenty more to investigate**: quantal chaos and the classical limit of quantum mechanics, friction and the transition to ground state, applications to astrophysics and cosmology and elementary particles.

But I want to close with one important yet rarely mentioned item: it's valuable to **develop your intuition** concerning quantum mechanics. We saw on page 39 that two common visualizations are flawed. Then on page 82 we found that *no* picture drawn with classical ink could successfully capture all aspects of quantum mechanics. How, then, can one develop a visualization or intuition for quantum mechanics? This is a lifelong journey which you have already begun. A good next step is to read the slim but profound book by Richard Feynman titled *QED: The Strange Theory of Light and Matter.*

The story of quantum mechanics began with the glowing logs of a campfire. It continued through atomic spectra; quantization, interference, and entanglement; Fourier sine series and partial differential equations. The story is not finished, and I invite you to add to it yourself.

Problem

8.1 **Questions** (recommended problem)
This is the end of the book, not the end of quantum mechanics. Write down any questions you have concerning quantum mechanics. Perhaps you will answer some of these through future study. Others might suggest future research directions for you.

Appendix A

Significant Figures

The impossibility of certainty

What is the pattern of thought most characteristic of science? Is it knowing that "net force causes acceleration rather than velocity"? Is it knowing that "momentum is conserved in the absence of external forces"? Is it knowing that "quantum mechanics has a classical limit"? No, it is none of these three facts — important though they are. The pattern of thought most characteristic of science is knowing that "every measurement is imperfect and thus every observed number comes with an uncertainty".

As with any facet of science, the proper approach to uncertainty is not "plug into a formula for error propagation" but instead "think about the issues involved". For example, in the course of building a tree house I measured a plank with a meter stick and found it to be 187.6 cm long. A more accurate measurement would of course provide a more accurate length: perhaps 187.64722031 cm. I don't know the plank's exact length, I only know an approximate value. In a math class, 187.6 cm means the same as 187.60000000 cm. But in a physics class, 187.6 cm means the same as 187.6??????? cm, where the question marks represent not zeros, but digits that you don't know. The digits that you *do* know are called "significant digits" or "significant figures".

This has an important philosophical consequence: Because all scientific conclusions are based on measurements, and all measurements contain some uncertainty, no scientific conclusion can be absolutely certain. All science is tentative; all those worshiping at the altar of science are fooling themselves.

This book is more concerned with the day-to-day practicalities of uncertainty than with the grand philosophical consequences. How do we express our lack of certainty? How do we work with (add, subtract, multiply, take logarithms of, and so forth) quantities that aren't certain?

Expressing uncertainty

The convention for expressing uncertain quantities is simple: any digit written down is a significant digit. A plank measured to the nearest millimeter has a length expressed as, say, 103.7 cm or 91.5 cm or 135.0 cm. Note particularly the trailing zero in 135.0 cm: this final digit is significant. The quantity "135.0 cm" is different from "135 cm". The former means "135.0?? cm", the latter means "135.??? cm". In the former, the digit in the tenths place is 0, while in the latter, the digit in the tenths place is unknown.

This convention gives rise to a problem for representing large numbers. Suppose the distance between two stakes is 45.6 meters. What is this distance expressed in centimeters? The answer 4560 cm is unsatisfactory, because the trailing zero is not significant and so, according to our rule, should not be written down. This quandary is resolved using exponential notation: 45.6 meters is the same as 4.56×10^3 cm. (This is, unfortunately, one of the world's most widely violated rules.)

Working with uncertainty

Addition and subtraction. I measured a plank with a meter stick and found it to be 187.6 cm long. Then I measured a dowel with a micrometer and found it to be 2.3405 cm in diameter. If I place the dowel next to the plank, how long is the dowel plus plank assembly? You might be tempted to say

```
  187.6
+   2.3405
  --------
  189.9405
```

But no! This is treating the unknown digits in 187.6 cm as if they were zeros, when in fact they're question marks. The proper way to perform the sum is

```
 187.6?????
+  2.3405??
 ----------
 189.9?????
```

So the correct answer, with only significant figures written down, is 189.9 cm.

Multiplication and division. The same question mark technique works for multiplication and division, too. For example, a board measuring 124.3 cm by 5.2 cm has an area given through

```
   1243?
  x  52?
--------
   ?????
  2486?
 6215?
--------
 64????
```

Adjusting the decimal point gives an answer of 6.4×10^2 cm^2

Although the question mark technique works, it's very tedious. (It's even more tedious for division.) Fortunately, the following rule of thumb works as well as the question mark technique and is a lot easier to apply:

> When multiplying or dividing two numbers, round the answer to the number of significant digits in the least certain of the two numbers.

For example, when multiplying a number with four significant digits by a number with two significant digits, the result should be rounded to two significant digits (as in the example above).

Rounding up vs. down. Mathematical operations result in an infinite number of digits, and they should be rounded to the appropriate number of significant digits. For example in the mathematical quotient

$$\frac{749.1}{152} = 4.928289\ldots$$

the result "4.928" should be rounded up to the physical result 4.93 with three significant digits. In contrast for the mathematical quotient

$$\frac{742.2}{152} = 4.882894\ldots$$

the result "4.882" should be rounded down to the physical result 4.88. But what about

$$\frac{731.9}{152} = 4.815131\ldots\ ?$$

When the leading non-significant digit is five, should it be rounded down to 4.81 or up to 4.82? It's permissible to round either way.

I was told in high school that when the leading non-significant digit was five, you could keep that digit, so that the result of the division above would be 4.815. I always loved it when the result came out this way: I was getting four significant digits for the price of three! You should not do this, because that fourth digit is in fact unknown. You're not getting four for the price of three, you're substituting fiction at the expense of fact.

Evaluating functions. How many significant figures does $\sin(87.2°)$ contain? We know that the real angle is somewhere between $87.200\ldots°$ and $87.300\ldots°$, so the real sine is somewhere between

$$\sin(87.200\ldots°) = 0.9988061\ldots$$
$$\sin(87.300\ldots°) = 0.9988898\ldots$$

The usual rule is to make the last significant digit in the result to be the first one from the left that changes when you repeat the calculation. In this case the first digit that changed was the "0" that changed to an "8", so we round the result to four significant figures, namely

$$\sin(87.2°) = 0.9988.$$

Numbers that are certain

Any measured number is uncertain, but *counted* and *defined* quantities can be certain. If there are seven people in a room, there are 7.0000000... people. There are never 7.00395 people in a room. And the inch is *defined* to be exactly 2.5400000... centimeters — there's no uncertainty in this conversion factor, either.

Conclusions

For most problems in this book the answer is an equation or a graph or a paragraph. But for a problem whose answer is a number, you must present that number with appropriate use of significant figures.

Appendix B

Dimensions

What does "dimensions" mean?

Suppose a table is two-hundred-thirty-six centimeters long or, in symbols,

$$\ell_T = 236 \text{ cm},$$

where ℓ_T represents the length of the table. This means that the table is 236 times as long as the length of the standard centimeter:

$$\ell_T = 236 \text{ cm} \quad \text{means} \quad \ell_T = 236 \times (\text{the length of the standard centimeter}).$$

In other words, the symbol "cm" used in the equations above represents "the length of the standard centimeter".

The symbol "ℓ_T" stands for "236 cm". That is, it stands for the number "236" times the length of the standard centimeter, or in other words, for the number "236" times the unit "cm". If I wrote "$\ell_T = 236$" instead of "$\ell_T = 236$ cm", I'd be dead wrong...just as wrong as if the solution to an algebra problem were "$y = 236\,x$" and I wrote "$y = 236$", or if the solution to an arithmetic problem were "236×7" and I wrote "236". In all three cases, my answer would be wrong because it omitted a factor. (These are, respectively, the factor of "the length of the standard centimeter", the factor of x, and the factor of 7.) The length of the table is not 236 — rather, the ratio of the length of the table to the length of the standard centimeter is 236.

Ignoring the units of a measurement results in practical as well as conceptual error. On 23 September 1999 the "Mars Climate Orbiter" spaceprobe crashed into the surface of Mars, dashing the hopes and dreams of dozens of scientists and resulting in the waste of $125 million. This spacecraft had survived perfectly the long and perilous trip from Earth to Mars. How could it have failed so spectacularly in the final phase of its journey? The manufacturer, Lockheed Martin Corporation, had told the the spacecraft controllers, at NASA's Jet Propulsion Laboratory, the thrust that the probe's rockets could produce. But the Lockheed engineers gave the thruster performance data in pounds (the English unit of force), *and they didn't specify which units they used.* The NASA controllers assumed that the data were in newtons (the SI unit of force).

Two Teams, Two Measures Equaled One Lost Spacecraft

By ANDREW POLLACK

LOS ANGELES, Sept. 30 — Simple confusion over whether measurements were metric or not led to the loss of a $125 million spacecraft last week as it approached Mars, the National Aeronautics and Space Administration said today.

An internal review team at NASA's Jet Propulsion Laboratory said in a preliminary conclusion that engineers at Lockheed Martin Corporation, which had built the spacecraft, specified certain measurements about the spacecraft's thrust in pounds, an English unit, but that NASA scientists thought the information was in the metric measurement of newtons.

The resulting miscalculation, undetected for months as the craft was designed, built and launched, meant the craft, the Mars Climate Orbiter, was off course by about 60 miles as it approached Mars.

"This is going to be the cautionary tale that is going to be embedded into introductions to the metric system in elementary school and high school and college physics till the end of time," said John Pike, director of space policy at the Federation of American Scientists in Washington.

Lockheed's reaction was equally blunt.

"The reaction is disbelief," said Noel Hinners, vice president for flight systems at Lockheed Martin Astronautics in Denver, Colo. "It can't be something that simple that could cause this to happen."

The finding was a major embarrassment for NASA, which said it was investigating how such a basic error could have gone through a mission's checks and balances.

"The real issue is not that the data was wrong," said Edward C. Stone, the director of the Jet Propulsion Laboratory in Pasadena, Calif., which was in charge of the mission. "The real issue is that our process

Continued on Page A16

New York Times, 1 October 1999, page 1, an embarrassing place to have your blunders published.

Modern information technology actually encourages mistakes like this. When you use a calculator, a spreadsheet, or a computer program, you enter pure numbers like "1.79", rather than quantities like "1.79 cm". So it's especially important to be on your guard and document your units when using computers. Keep the units in your mind, even if you can't keep them in your calculator!

A nitpicky distinction is that the length of the table has the *units* of "centimeters" and the *dimensions* of "length". If the length of the table were measured in centimeters or meters or even in cubits it would still have the dimensions of length. But in everyday language the terms "units" and "dimensions" are often used interchangeably.

Unit conversions

It is standard usage to refer to the length of the standard centimeter by the symbol "cm". But in this appendix I'll also refer to the length of the standard centimeter by the symbol ℓ_{cm}. Similarly I'll call the length of the standard meter either "m" or ℓ_m.

You know that if a table is 236 centimeters long it is also 2.36 meters long:

$$\ell_T = 236 \text{ cm} = 236\,\ell_{cm} = 2.36 \text{ m} = 2.36\,\ell_m.$$

How can this be? It's certainly *not* true that $236 = 2.36$! It's true instead that $236 \text{ cm} = 2.36 \text{ m}$ because the length of a meterstick is 100 times the length of a standard centimeter:

$$\ell_m = 100\,\ell_{cm}.$$

This tells us that

$$2.36 \text{ m} = 2.36\,\ell_m = 2.36 \times (100\,\ell_{cm}) = 236\,\ell_{cm} = 236 \text{ cm},$$

or, in the opposite direction,

$$\begin{aligned} 236 \text{ cm} &= 236\,\ell_{cm} = 236\,\ell_{cm}(1) = 236\,\ell_{cm}\left(\frac{\ell_m}{100\,\ell_{cm}}\right) \\ &= 236\,\cancel{\ell_{cm}}\left(\frac{\ell_m}{100\,\cancel{\ell_{cm}}}\right) = 2.36\,\ell_m = 2.36 \text{ m}. \end{aligned}$$

In short, the symbol "cm" can be manipulated exactly like the symbol "ℓ_{cm}", because that's exactly what it means!

Incompatible dimensions

If I walk for 4.00 m, and then for 59 cm, how far did I go? The answer is 459 cm or 4.59 m, but not $4.00 + 59 = 63$ of anything!

If I walk for 4.00 m, and then pause for 42 seconds, how far did I go? Certainly *not* 4.00 m + 42 s. The number 46 has no significance in this problem. For example, it can't be converted into minutes.

In general, *you can't add quantities with different units.*

This rule can be quite helpful. Suppose you're working a problem that involves a speed v and a time t, and you're asked to find a distance d. Someone approaches you and whispers: "Here's a hint: use the equation $d = vt + \frac{1}{2}vt^2$." You know that this hint is wrong: The quantity vt has the dimensions of [length], but the quantity $\frac{1}{2}vt^2$ has the dimensions of [length]×[time]. You can't add a quantity with the dimensions of [length]×[time] to a quantity with the dimensions of [length], any more than you could add 42 seconds to 4.00 meters.

A famous use of dimensional analysis

Dimensional analysis sometimes makes it possible to uncover a lot about complicated situations if you can only ferret out the essential features of the situation. For example the fluid flow expert G.I. Taylor was able to deduce the explosive yield of the first nuclear bomb from a sequence of photographs of the expanding fireball published in *Life* magazine.[2]

[2]Details presented in Michael B.A. Deakin, "G.I. Taylor and the Trinity test" *International Journal of Mathematical Education in Science and Technology* **42** (2011) 1069–1079.

Taylor realized that this sequence of photos showed a shock wave expanding into an undisturbed medium, and he knew from his previous experience that the radius of the fireball r could depend upon only three things: the density of undisturbed air ρ, the energy released through the explosion E, and the time since the explosion t.

quantity	dimensions
ρ	[mass]/[length]^3
E	[mass] $\times$ [length]^2/[time]^2
t	[time]
r	[length]

Taylor knew that to build r out of ρ, E, and t, he had to cancel out the [mass] that appears in ρ and E but that cannot enter into r. Thus he had to build r out of E/ρ and t.

quantity	dimensions
E/ρ	[length]^5/[time]^2
t	[time]
r	[length]

Now Taylor had to cancel the [time]^2 from the denominator of E/ρ using the variable t:

quantity	dimensions
Et^2/ρ	[length]^5
r	[length]

Taylor concluded that

$$r(t) = C\sqrt[5]{Et^2/\rho}$$

where C is some dimensionless constant like π or 7 but *not* a number with dimensions like 9.8 m/s^2.

Sure enough, a plot of r as a function of $t^{2/5}$, with data taken from the magazine photo sequence, yielded a straight line. The energy released by the explosion was

$$E = \frac{1}{C^5}\rho\frac{r^5}{t^2}.$$

Taylor had experimental data suggesting that $1/C^5 = 1.033$, and from this he was able to find the energy output of the nuclear bomb explosion at a time when this precise number was a closely guarded secret.

Conclusions

For most problems in this book the answer is an equation or a graph or a paragraph. But for a problem whose answer is a number, you must present that number with units attached.

Problems

B.1 **A new law of nature?**

It has been proposed that the speed of sound v_s and the speed of light c are related through $v_s = \frac{1}{2}\sqrt[3]{c}$. Check the accuracy of this formula using speeds expressed in meters/second, then recheck its accuracy using speeds expressed in kilometers/second. (According to the book *U.S. Standard Atmosphere, 1976*, the standard speed of sound at sea level is 340.29 m/s.) Is this proposal a new and surprising law of nature, or merely a coincidence? Explain.

B.2 **Sound speed**

The speed of sound v_s in air, in a given room, could reasonably depend on three things: the air density ρ, the air pressure p, and the room volume V. In other words

$$v_s = C\rho^x p^y V^z$$

where C is some dimensionless number. Find the exponents x, y, and z.

B.3 **Physics in film**

Alfred Hitchcock's 1935 film *The 39 Steps* is one of the great spy thrillers of all time. In the film, several men and women crisscross England and Scotland in pursuit of an important but unspecified secret document. Only in the final minute of the film does the audience find that the the document contains the specifications for a completely silent aircraft engine, and that these specifications hinge upon "the secret formula

$$\left(r - \frac{1}{r}\right)^{\gamma}$$

where r represents the ratio of compression and γ the axis of the fluid line of the cylinder."

Show that this formula is not worth the pursuit of a cadre of spies, and in fact is entirely without meaning.

Clue: You know that 5^3 means "5 times itself three times" or $5^3 = 5 \times 5 \times 5$. What does $5^{(3 \text{ feet})}$ mean? How is it related to $5^{(1 \text{ yard})}$? [[Note for non-Americans: The "foot" and the "yard" are archaic units of length used in the United States of America. A foot is about 30 cm long, and a yard is defined as exactly three feet.]]

Appendix C

Complex Arithmetic

The familiar numbers like 17, $\frac{3}{5}$, $\sqrt{2}$, $-\pi$ and so forth are called "real numbers". The square of any real number is non-negative. But we can imagine a new category of numbers that have negative squares. We first imagine the number i, with $i^2 = -1$. Then we can imagine the number 3 times i, with $(3i)^2 = -9$. These are called "imaginary numbers".

The names "real" and "imaginary" are unfortunate. Numbers are useful abstractions that exist in our minds: you've seen two hands, and two fingers, and two apples; you've seen the Arabic numeral "2" and the Roman numeral "II", which are made of ink; you've seen the English word "two", the German word "zwei", and the Somali word "laba", again made of ink; but you've never seen the number 2, which is made of pure thought. In the usual sense of the words "real" and "imaginary", no number is real; all numbers are imaginary.

The sum of a real number and an imaginary number is called a "complex number" (another unfortunate name). Just as the real number x can be profitably visualized as a point on the one-dimensional real line, so the complex number $z = x+iy$ can be profitably visualized as the point (x, y) on the two-dimensional "complex plane". This book assumes some background in complex arithmetic. If your knowledge is rusty, these problem should grease your mental gears.

Problems

C.1 **Complex sum and product**

Find the sum $(2+3i)+(-3+5i)$, the product $(5+7i)(2-i)$, and the square $(3+i)^2$.

C.2 **Cartesian and polar forms of a complex number**

The "Cartesian form" of a complex number z is $x+iy$, where x and y are real. (The quantity x is called the "real part of z": $x = \Re e\{z\}$, while the quantity y is called the "imaginary part of z": $y = \Im m\{z\}$.) The "polar form" of a complex number is $re^{i\theta}$, where r and θ are real and r is non-negative. (The quantity r is called the "magnitude", while the quantity θ is called the "phase".) Using the Euler relation $e^{i\theta} = \cos\theta + i\sin\theta$, show that these forms are related through

$$r = \sqrt{x^2+y^2} \quad \text{and} \quad \tan\theta = y/x.$$

C.3 **Express in polar form**

Express in polar form: $2+\sqrt{12}\,i$ and $-1+\sqrt{3}\,i$.

C.4 **Multiplication of complex numbers**

Find the product $(2+\sqrt{12}\,i)(-1+\sqrt{3}\,i)$ using both Cartesian and polar forms.

C.5 **Polar form of i and 1**

Show that $i = e^{i(\frac{1}{2}\pi+2\pi n)}$, where n is any integer (positive, zero, or negative). Similarly, find an infinite number of polar representations of the number 1.

C.6 **Complex conjugate**

If $z = x+iy$, where x and y are real, then the "complex conjugate" of z is defined as $z^* = x - iy$. Show that $r^2 = zz^*$. (The magnitude r is also called $|z|$, so this result is often written $|z|^2 = zz^*$.)

Appendix D

Problem-Solving Tips and Techniques

A physicist can wax eloquent about concepts like interference and entanglement, but can also use those concepts to solve problems about the behavior of nature and the results of experiments. This appendix gives general advice on problem solving, then lists the problem-solving tools introduced and elaborated upon in this book.

You have heard that "practice makes perfect", but in fact practice makes permanent. If you practice slouchy posture, sloppy reasoning, or inefficient problem-solving technique, these bad habits will become second nature to you. For proof of this, just consider the career of ⟦insert here the name of your least favorite public figure, current or historical, foreign or domestic⟧. So I urge you to start now with straight posture, dexterous reasoning, and facile problem-solving technique, lest you end up like ⟦insert same name here⟧.

An approach to problem solving

Suppose you want to travel from San Diego to Boston. You start by deciding whether to fly, take a bus, or drive a car. If you decide to fly, you then make subsidiary decisions like choice of airline. If you decide to drive, you make different subsidiary decisions: Should you first change your car's oil? Should you take a side trip to the Grand Canyon? Or to visit your friends in Boulder, Colorado? What you *don't* do is just step out of your front door and walk northeast: you make a plan before taking that very first step.

And just as your journey begins before you take your first step, so it extends after you take your last step. Any journey, properly considered, includes reflection upon that journey. This might be merely technical ("I'll never travel on that bus line again.") or it might open a door to future travel ("The Grand Canyon was so spectacular! Next time I'll hike from the rim down to the Colorado River.") or it might be deeper still ("My friends in Boulder seem so happy together. I need to rethink my plan of remaining single all my life.").

As with travel, so with physics problem solving. The first step is to **understand the problem**. What is given? What is asked for? For problems in classical mechanics and electromagnetism, one tool for understanding the problem is to **sketch the situation**. In relativity, if often helps to make two sketches: one for each reference frame. Some quantum mechanics problems are so abstract that a sketch doesn't help you understand the problem, but often doodling plays that same role.

The next step is to **select a strategy** — a key idea to employ — before rushing in to make detailed derivations. Is this a time evolution problem? An energy eigenproblem? An interference or entanglement problem?

Once you pass on to implementing your strategy, **keep your goal in mind** to avoid deriving endless numbers of equations that are true but that don't help you reach your goal.

Finally, once you're reached that goal, **reflect on your final result**. What is nature trying to tell you through this problem? Is the result in accord with your expectations? A good example of this stage of problem solving appears on page 171. Instead of reaching equation (4.105), saying "That's the end", and heading to bed for some well-earned sleep, we spent two paragraphs on the consequences of that equation, found them remarkable and unexpected, and used them to illuminate the role of interference in quantum mechanics. I.I. Rabi reflected on the consequences even more deeply, and used those reflections to invent the atomic clock. Another example concerns the Planck radiation law (1.13): pages 16–23, plus sample problem 1.2.1, are ten pages of reflection upon that single brief equation.

As with travel, such reflection might be merely technical ("Why did I have to work out that integral in detail? I should have seen from symmetry that the result would be zero.") or it might be deeper ("When Styer said an atom might not have a position, I thought he was spouting bullshit. But after working problem 2.7, 'Bomb-testing interferometer', I realize that I have to rethink my ideas about how atoms behave.").

List of problem-solving tools

Appendix E

Catalog of Misconceptions

Effective teaching does not merely instruct on what is correct — it also guards against beliefs that are not correct. There are a number of prevalent misconceptions concerning quantum mechanics. This catalog presents misconceptions mentioned in this book, together with the page number where that misconception is pointed out and corrected.

Index

CPSIA information can be obtained
at www.ICGtesting.com
Printed in the USA
BVHW010302040422
633154BV00001B/1